# INTRODUCTION TO
# Statistical Methods

## Second Edition

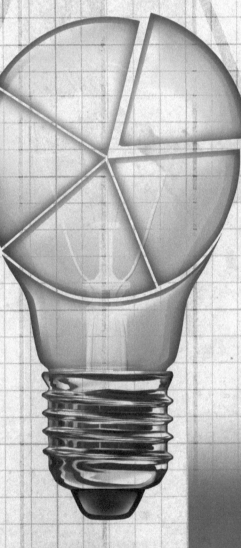

# Kendall Hunt
publishing company

Amy B. Maddox

Cover image © Shutterstock, Inc.

publishing company

www.kendallhunt.com
*Send all inquiries to:*
4050 Westmark Drive
Dubuque, IA  52004-1840

# CONTENTS

# INTRODUCTION TO STATISTICS

## ➡ Section 1: Introduction:

What exactly is statistics? *Statistics is the art of collecting, organizing, summarizing, and describing data as well as drawing inferences from the data.*

- The goal of statistics is to draw conclusions about the real world based on observations one makes. In the language of statistics, the goal is to draw conclusions about a population based on what is observed from a sample taken from that population. This might sound simple, but due to the fact that many times, there is more than one phenomenon occurring, this can be quite a challenge.

- Consider the following example courtesy of Dr. John Seaman. Suppose you are concerned about the difficulties some couples have in conceiving a child.

  - It is thought that women exposed to a particular toxin in their workplace have greater difficulty becoming pregnant compared with women who are not exposed to the toxin. You conduct a study of such women recording the time it takes to conceive.

  - Of course, there is natural variability in the time it takes to achieve a pregnancy, which is attributable to many causes aside from the toxin. Nevertheless, suppose you finally determine that those females with the greatest exposure to the toxin had the most difficulty getting pregnant.

  - But, what if there is a variable you did not consider that could be the cause? No study can consider every possibility.

  - For example, it turns out that women who smoke while they are pregnant reduce the chance their daughters will be able to conceive because the toxins involved in smoking effect the eggs in the female fetus!

  - If you didn't record whether or not the females had mothers who smoked when they were pregnant, you may draw the wrong conclusion about the industrial toxin.

- In statistics, the conclusions we make always come with some amount of uncertainty. i.e., we never really know if we made the correct decision. This is because we are dealing with a sample and not the entire population, i.e., we are not able to measure every woman's time to conception, only a sample of women.

- In statistics we quantify our uncertainty about our decision using probability, which we will discuss later. This allows us as statisticians to determine how "good" our decisions are.

Consider the following illustration: We have a specific population that we would like to know something about, so we randomly sample from that population, look at a picture of the sample, and then, with the idea of repeated sampling, create a population model that we use to draw conclusions about the population.

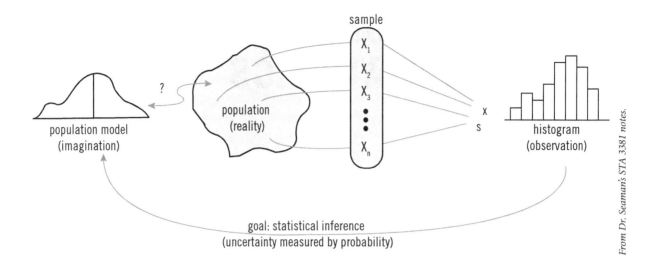

We can divide the study of statistics into two broad sections, descriptive and inferential. In descriptive statistics one is simply describing what they observed from the sample, whereas in inferential statistics, one is using the information discovered in the sample to draw conclusions about the population.

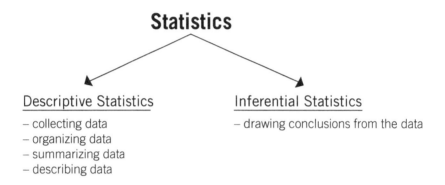

We will begin discussing the areas of descriptive statistics, but will spend the majority of the class in inferential statistics. I like to think of the notes in the beginning of the semester as giving us the "tools" we need to do inferential statistics. Obviously before you can begin drawing conclusions, you have to be able to collect and organize the data and understand the data. But first, we must learn some basic definitions used in statistics.

# Definitions

- *Population* – the complete set of units we are interested in studying. The population may be people, animals, or objects. Most people tend to think of a population as containing people, but if I am interested in studying tires, then my population is ALL tires. (sometimes referred to as *Target Population*.)

  It can sometimes be a challenge to determine the population. Specifying the population is the first and one of the most important steps of statistics.

  The size of the population is denoted by the symbol, $N$.

  (Statistics is a language that uses symbols and notations to convey meaning. It is very important that you learn the "language" of statistics by learning the symbols and the correct notation)

Example: What is the Population in a political poll?

**A.** All eligible voters

**B.** All registered voters

**C.** All persons who voted in the last election

Answer: The answer is B. If we want to know who is going to win, our population is only those registered to vote because they are the only ones that can determine the outcome.

- *Sample* – a subset of the population.
- *Well Representative Sample* – a sample in which the characteristics of the sample match the characteristics of the population. (Our goal is to always have a well-represented sample, but as we will see, that is sometimes a challenging goal.)

The sample size is denoted by the symbol, $n$.

- *Parameter* – a numerical value based on the population. Unless we have information from every element of the population, the parameter is unknown.
- *Statistic* – a numerical value based on the sample. Using the information found in the sample, we can calculate the statistic.

  **In drawing conclusions, we use the statistic to draw conclusions about the parameter.** We will learn later how to do this, but this is a very important concept to understand!

- *Variable* – a characteristic that takes on different values for different people, places, or things is labeled a variable. For a value to be a variable, it must potentially vary for all elements of the population.
- *Constant* – values in the population that do not vary are known as constants.

  - Is a parameter a variable or a constant?

    Answer:

  - Is a statistic a variable or a constant?

    Answer:

- *Probability* – measures how likely it is for something to occur. We can only calculate probabilities about variables, not constants.

  - Can we calculate probability about a parameter?

    Answer:

- Can we calculate probability about a statistic?

  Answer:

- *Random Variable* – A variable is a *random variable* if the exact value of the variable cannot be predicted in advance. This occurs when the value of a variable is obtained as a result of a chance factor.

- There are two "types" of random variables:

  - *Quantitative Random Variable* – a variable that can be measured in the usual sense.

  - *Qualitative Random Variable* – a characteristic that cannot be measured in the usual sense; it can only be put into categories. (Also called a *Categorical Variable*.)

- There are two types of Quantitative Random Variables:

  - *Discrete Random Variable* – when the variable can only assume specific values.

  - *Continuous Random Variable* – when the variable can theoretically assume any value on a given interval.

    Example: Suppose my area of research concerns feet, specifically the size of my students' feet. What are two questions/things that I can ask or do to determine the size of your feet?

    Answer 1.

    Is _____ discrete or continuous?

    Answer 2.

    Is _____ discrete or continuous?

- Here is a flow chart showing the different types of random variables. It is important to understand the different types of random variables. The type of random variable will determine the method used to analyze the data.

# Measurement and Measurement Scales

- *Measurement* – the assignment of numbers to objects or events according to a set of rules. There are four measurement scales that result from the fact that measurements may be carried out using different rules.

  - *The Nominal Scale* – "naming observations" or classifying them into various mutually exclusive categories, for example, male/female.

  - *The Ordinal Scale* – when the measurements can be ranked according to some criterion, for example, intelligence– low/average/high (Note that arithmetic operations are not meaningful.)

  - *The Interval Scale* – when the measurements cannot only be ranked, but the distance between any two measurements is known. There is no true zero point. For example, temperature using Celsius or Fahrenheit scale. 60°F is 30°F greater than 30°F, but 60°F is not twice as hot as 30°F.

  - *The Ratio Scale* – measurements in which equality of ratios as well as equality of intervals is known. There is a true zero point. For example, height, weight, length, time to complete a task. Ten minutes is twice as long as five minutes.

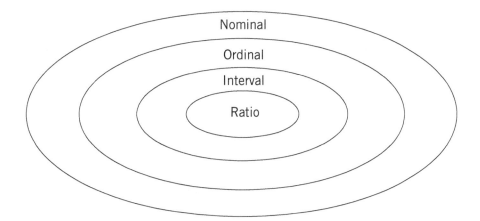

## ➡ Section 2: Descriptive Statistics – Collecting Data:

This is a very important area of statistics. If you have bad data, then no matter how great your techniques are, your results are going to be meaningless. We will briefly discuss some general techniques for collecting data, but we will by no means cover all the different ways you can collect data. This is just a small overview of the topic.

- Methods for Obtaining Data

  - *Census*

  - *Sampling*

  - *Experimentation*

- Census: A *census* occurs when information is taken from ALL subjects/objects of the population.

  In other words, when taking a census, you are not collecting a sample; you are collecting information from EVERY element of the population.

- Any sample can be classified as one of the following:

  1. Nonprobability Samples

  2. Probability Samples

- *Nonprobability Samples* – the laws of probability are NOT used to obtain the samples.

  Listed below are some examples of nonprobability samples. This is by no means a complete listing of all different types of nonprobability samples, but it is a list of the ones more commonly used.

  - *Sample of Convenience* – the samples already exist.

    For example, since I teach a pre-med class, if my population is pre-med students, then I could use my class as a sample. This would be a sample of convenience.

  - *Haphazard Selection* – the researcher and the subjects casually meet.

    For example, if my population is shoppers at a suburban mall and I go to the mall and just casually approach people to survey them, then I have a haphazard selection.

  - *Judgment Sampling* – the researcher uses his/her own judgment in choosing the subjects.

    For example, suppose again my population is shoppers at a suburban mall and I go to the mall with the goal of surveying shoppers to determine their opinions on the new shopping hours. If I purposely only approach those whom appear to be "serious" shoppers and not just using the mall for exercise or entertainment, then I have used judgment sampling.

  - *Expert Sampling* – an "expert" picks the subjects.

    For example, if, as the researcher, my population consists of dolphins and I let the biologist choose the sample, then I have an expert sample because the "expert" about dolphins chose the sample.

  - *Quota Sampling* – subjects are chosen so as to satisfy certain quotas that match the population.

    A quota sample can be a probability sample if the subjects are randomly chosen, but I included it here because this is generally how you see a quota sample done.

    For example, suppose my population of Baylor students contains 22% freshmen, 20% sophomores, 19% juniors, 18% seniors, and 21% graduate students. If I choose my sample so that 22% of the sample are freshmen, 20% are sophomores, 19% are juniors, 18% are seniors, and 21% are graduate students, then I have a quota sample.

- Probability Samples – the sample is obtained by a chance process. In other words, the laws of probability are used to choose the sample.

  - *Simple Random Samples* – all samples of the same size have equal probability of being selected.

    One good analogy of obtaining a simple random sample is as follows: Suppose I want to randomly sample ten of my nursing students. To do this, I can put all their names in a sack, shake the sack, pull out a name, shake the sack again, pull out another name and continue this process until ten names have been drawn. Now in reality, we use a computer to randomly generate a sample.

    Question: Do you think it is possible to get a sample of 100 randomly chosen Baylor students that does not contain any seniors?

    Answer: Yes, it is possible. Not all random samples are well-representative samples of the population. When appropriate we can improve the sample by performing one of the sampling schemes below.

- *Stratified Random Samples* – population is divided into subpopulations (called strata) and a random sample is obtained from each strata.

  For example, suppose I want to obtain a sample of health education majors. To insure that I get all student classifications in my sample, I can divide my population of health education majors into five stratas: freshman, sophomore, junior, senior, and graduate student. Then I would take a random sample from each stratum to obtain a stratified random sample.

- *Cluster Samples* – population is once again divided into stratas or clusters. The researcher randomly selects some of the strata or clusters and selects ALL subjects from the chosen stratas. (Also when the subjects are naturally clustered together).

  For example, suppose I want to sample residents of a very large apartment complex that has units that are broken into "groups" (clusters) of six apartments. To obtain a cluster sample, I would randomly choose the "groups" (clusters) to sample and then sample every apartment unit in that chosen group.

  Another example: Suppose my population is tires. Since tires are clustered on vehicles, I would randomly choose the car and then sample each tire on the car.

- *Systematic Sampling* – choose a random starting point from a list of subjects and then select every $n^{\text{th}}$ subject on the list.

  For example, suppose I want to sample ten patients at General Mercy Hospital. Suppose the administrator at General Mercy Hospital has a numbered list of ALL 100 patients currently in the hospital. If I randomly choose a starting number between one and ten, say three, and then choose every tenth one, the third, thirteenth, twenty-third, etc., then I have a systematic sample.

- Sampling Error: the extent to which the value of a statistic may differ from the parameter it predicts because of the way results vary from sample to sample. For example, suppose the true value of the parameter is 56%, but we predict 53%. Then the sampling error is -3%. (Gallop poll example.)

- Biases can cause a researcher to draw incorrect conclusions!

  - *Selection Bias* (also called undercoverage) occurs when a specific characteristic or segment of the population is excluded.

    For example, if my population is all students at the local university and I only go to the library to survey students, then my sample will contain selection bias because I excluded the students that do not study at the library.

  - *Response Bias* occurs when subjects respond incorrectly by lying or exaggerating or forgetting or not understanding the question. The question may also be a misleading question.

    For example, if you ask individuals if they have ever taken illegal drugs, they may not admit to doing so.

  - *Nonresponse Bias* occurs when the response rate is low.

    The type of people who will respond to a survey is very different from the type of people who will not respond; that difference can significantly affect the results of the survey.

- There are two "types" of Experiments:

  1. Observational Study – simply observing the subjects in a specific situation or "natural habitat" or when simply comparing two or more groups. An observational study can only show association, not causation. In other words, you cannot show causation with an observational study!

**2.** Designed Experiment – has three basic principles:

   **a.** Control

   **b.** Randomization – randomly divides subjects into groups

   **c.** Replication – have sufficient number of subjects used to have groups as similar as possible in all aspects, except the treatment

- *Treatment Group* – the group that receives the treatment.

- *Control Group* – group that does not receive the treatment.

- *Confounding factor* – anything that affects the outcome of the experiment other than the treatment. Also called an extraneous factor.

- *Blind Study* – when the subjects don't know if they are receiving the treatment or a placebo

- *Double-Blind Study* – when neither the subjects nor the doctors know if subjects are in the treatment or control group.

- The four basic experimental designs

   - Completely Randomized Design

   - Randomized Block Design

   - Latin Square Design

   - Factorial Design

- Problems with Experiments

   - *Placebo Effect* – when subjects improve because they believe they are receiving the treatment

   - *Rosenthall Effect* – when the researcher or experimenter unintentionally influences the outcome through facial expressions, body language, or voice.

   - *Hawthorne Effect* – when the outcome is affected because people change how they behave because they know they are being watched.

   - *Missing Data*

# ➡ Section 3: Descriptive Statistics – Organizing Data:

Once the data has been collected, it is important to organize the data in a table or graph. The goal of organizing data is to make the data more "comprehendible." If you simply list out the data in rows of numbers, then there is little meaning to the data; it is simply a row of numbers. So we group the data into tables or make graphs (a picture) of the data. The type of graph created varies depending upon if the data is quantitative or qualitative. For quantitative data, the type of table or graph is also dependent upon if the data is discrete or continuous. Some common tables and graphs are listed below.

| Tables | Graphs |
|---|---|
| Frequency Distribution | Dot Plot |
| Grouped Frequency Distribution | Histogram |
| Relative Frequency Distribution | Bar Graph |
| | Pie Chart |

- The following tables are typically used for quantitative data:
  - Frequency Distribution: a table that lists the data values and the number of times the data value occurs. (The number of times a data value occurs is called its frequency.) This table can be used for discrete or continuous data.

    Example: Suppose the following frequency distribution table represents the age at which a child sampled first loses a tooth.

| X | 3 | 4 | 5 | 6 | 7 |
|---|---|---|---|---|---|
| f(x) | 1 | 5 | 10 | 6 | 3 |

  - The symbol, $X$, is used to represent the unique data values.
  - The notation, $f(x)$, is used to represent the frequency of the data.
  - How many children sampled were five years old when they first lost a tooth?

    Answer:


  - What is the value $n$?

    Answer:


- Dot Plot: A "picture" of a frequency distribution commonly used is the Dot Plot. In a Dot Plot, dots are used to represent the frequency of the data.

  Example:

- Relative Frequency Distribution: a table that lists the data values and the percentage of times the data value occurs. (The percentage of times a data value occurs is called its *relative frequency*).

  Example: Consider again the age at which a sampled child first loses a tooth. The frequency is now listed as a percentage instead of as a count.

| X | 3 | 4 | 5 | 6 | 7 |
|---|---|---|---|---|---|
| f(x) | 0.04 | 0.20 | 0.40 | ? | ? |

  - Again, the symbol, $X$, is used to represent the unique data values.
  - The notation, $f(x)$, is now used to represent the relative frequency of the data.

    What is the relative frequency for the data value, $X = 6$?

    Answer:

    What is the relative frequency for the data value, $X = 7$?

    Answer:

    What is the sum of the relative frequencies?

    Answer:

- Grouped Frequency Distribution. When the sample size is large, the data is typically combined into groups. Recall that the goal of organizing the data into tables and graphs is to aid in understanding the meaning of the data. If the frequency distribution becomes too large, the "meaning" of the data is difficult to "see," thus the data is combined into groups.

  The groups can be determined by "natural" grouping in the data or by the sample size, $n$. For example, if the data obtained is grades, these are naturally grouped in terms of As, Bs, etc. If the data does not naturally fall into groups, the size of the data set, $n$, determines the number of groups used. A good rule of thumb for determining the number of groups in a grouped frequency distribution is $\sqrt{n}$. A requirement of a grouped frequency distribution is that every data value should belong to a group and the first and last group should contain a data value.

  (I am not going to spend time teaching you how to create a grouped frequency distribution by hand. The main thing is to understand what a grouped frequency distribution is and how to read one.)

- Histogram – a "picture" of a grouped frequency distribution is called a histogram. The horizontal axis represents the grouped data. The height of the bar drawn represents the frequency or relative frequency of the data. The width of the bar equals the width of the group. For example, if grouping by grade, the points 90 – 100 represent an A, so the width of the bar equals ten. For data that is not grouped, the width of the bar is determined by the number of data values and the number of groups chosen.

(Again, I am not going to spend time in class teaching you how to create a histogram by hand. We will use technology to create histograms.)

Example: Suppose again we are interested in the age at which a child first loses a tooth. To be measured more exactly, the data set takes into account a more exact age. Instead of just putting 5 for five years, it includes the months as well. For example, the value 5.25 implies the child was five years and three months old.

---

We will use JMP to create the histogram for us. Listed below are the steps:

1. Under file, choose New, then choose Data Table. Click on the first cell of column 1 to enter the unique data values. Choose New Column to get column 2. Click on the first cell of column 2 to enter the frequency of the data.

2. Under the analyze menu, choose distribution. Click on the column containing the unique data and click on the Y, Columns button. Click on the column containing the frequencies and click on the Freq button. Click Ok.

3. Under the red triangle, choose histogram to change the histogram.

4. Choose show counts to show the counts or choose show percents to get the percentages.

5. To add the label, place cursor on the x-axis and right click, choose add axis level.

6. Once a label is added, you can edit it by clicking on it.

---

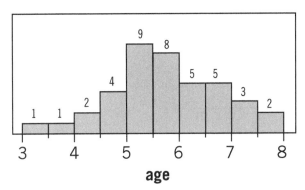

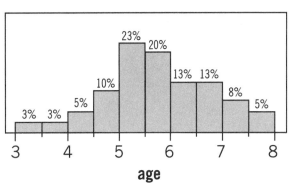

- Another graph used for quantitative data, discrete or continuous, is called a stem-n-leaf plot: To get the stem-n-leaf plot in JMP, simply choose it as a choice under the options.

| Stem | Leaf | Count |
|---|---|---|
| 7 | 56 | 2 |
| 7 | 122 | 3 |
| 6 | 56689 | 5 |
| 6 | 22333 | 5 |
| 5 | 56788889 | 8 |
| 5 | 122233344 | 9 |
| 4 | 8899 | 4 |
| 4 | 13 | 2 |
| 3 | 8 | 1 |
| 3 | 1 | 1 |

3|1 represents 3.1

- The following graphs are typically used for qualitative data:

  - Bar graph: In a bar graph, a bar is used to represent each "category of the data." The height of the bar represents the total percentage or the total count of how many data values fall in the specific category. With a bar graph, the bars are not touching.

To create the bar graph using JMP use the following steps:

1. Once the data is inputed, under the graph menu, choose chart.

2. Click on the column containing the unique data and click on Categories, X level. Click on the column containing the frequencies and click the Freq button under the Additional Roles section.

3. To obtain the percentages, under the red triangle by the word chart, choose Label Options and then choose Label by percent of Total Value.

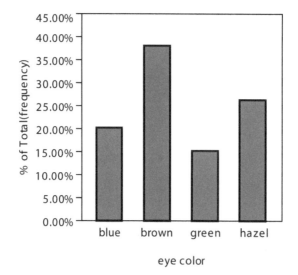

- Pie chart: In a pie chart, each category is represented by a slice of the pie. The size of the pie slice represents the total percentage for each category. For a pie chart, we can change the color of the slices to represent the categories.

> To create the pie chart using JMP use the following steps:
>
> 1. Once the data is inputed, under the graph menu, choose chart.
>
> 2. Click on the column containing the category and click on Categories, X level. Click on the column containing the frequencies and click the Freq button under the Additional Roles section.
>
> 3. On the left side, there is an options choice. Change this from bar graph to pie chart.
>
> 4. To change the color of the slice. Put cursor on the color box in the legend under the pie chart, right click, and choose color.

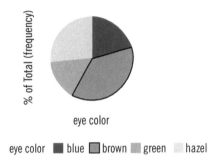

eye color

eye color ■ blue ■ brown ■ green ■ hazel

# ➡ Section 4: Descriptive Statistics – Describing Data

When describing the data, the goal is to calculate a single value to describe a certain aspect of the data. There are three different aspects of the data we like to describe: the center, the variability, and the location.

Measures of Central Tendency – conveys a "typical" value of the data set. There are three measures of central tendency:

1. Mean – the center of the data

2. Median – the middle of the data

3. Mode – the data value that occurs the most often

The Mean

- The mean is the true center of the data. It is the balancing point of the data. Graphically, it is the point at which the graph would be balanced. The deviation above the mean will always equal with the deviation below the mean.

  Let $X$ = random variable.

There are two different means, the population mean and the sample mean:

$$\mu = \frac{\sum_{i=1}^{N} X_i}{N}$$    $\mu$ is the population mean

- to find $\mu$ add up all the data values in the population and divide by $N$.

Is $\mu$ a parameter or a statistic?

Is $\mu$ a constant or a random variable?

$\mu$ is _____ unless a _____ is performed.

Can we calculate probabilities with respect to $\mu$?
Answer:

$$\bar{x} = \frac{\sum_{i=1}^{n} X_i}{n}$$    $\bar{x}$ is the sample mean

- to find $\bar{x}$ add up all the data values in the sample and divide by $n$.

Is $\bar{x}$ a parameter or a statistic?

Is $\bar{x}$ a constant or a random variable?

$\bar{x}$ is _____. It is calculated from the _____.

Can we calculate probabilities with respect to $\bar{x}$?

Answer:

**\*\* We use $\bar{x}$ as an estimate for μ.**

- Properties of the mean:
  - Most commonly used.
  - May not be a value in the data set.
  - Is not a resistance measure, which is a disadvantage to the mean. (We will talk about this more later.)
  - We do not round the mean to a whole number, even if the element in the population can't be a whole number. For example, the average number of kids per family is 3.2. Even though you cannot have 0.2 of a child, we don't round the mean to 3; we keep it as 3.2.

The Median

- The median is the value such that 50% of the data values are less than that value and 50% of the data values are greater than that value. The median divides the data into half.
- **Notation**: M = the population median

  $\tilde{x}$ = the sample median

To find:

1. Rank the data (i.e., put the data in order)
2. Find the location of the data using the formula $(n+1)/2$.

   If there are an even number of data values, then the location value above will not be an integer. So to find the median, you average the two middle values.

   For example, if the data set is

   $$2 \qquad 6 \qquad 8 \qquad 10$$

   Then location of the median is $(4 + 1)/2 = 2.5$, and thus the median is 7

   If there are an odd number of data values, then the location value above will be an integer. So to find the median you just determine the data value in that location.

   For example: if the data set is

   $$2 \qquad 8 \qquad 10 \qquad 14 \qquad 18$$

   Then the location of the median is $(5 + 1)/2 = 3$, and thus the median is 10.

- Properties of the median:
  - May not be a value in the data set.
  - Is a resistance measure—which is an advantage of the median.

# Definitions

*Outlier* – a data value that is an extreme value

*Resistance Measure* – a measure that is not affected by outliers

- Example: Consider the following data set

$$46 \quad 44 \quad 43 \quad 38 \quad 45$$
$$42 \quad 45 \quad 44 \quad 46 \quad 44 \quad 135$$

- Note that the sample mean is $\bar{x} = 52$ and the sample median is $\tilde{x} = 44$.

Which measure do you think best represents the typical value in the data set?

Answer:



Why is there such a difference between the mean and the median?

Answer:



Suppose the data set above represents the average starting salaries (in thousands) for nurses.

If, as the researcher, I report that the average starting salary is 52,000, am I correct?

Answer:



Am I being misleading?

Answer:



It would be better for me to report that the _____ starting salary is _____.

When the data set contains outliers, the _____ should be used as the measure of central tendency.

More on Outliers:

Recall that outliers are data points that are much larger or much smaller than the other data values, i.e., an extreme value. However, the "extremes" are not necessarily always outliers. There are different methods for determining if an extreme value is an actual outlier. We will discuss these later.

Outliers can sometimes lead to complications in data analysis and erroneous conclusions if not treated appropriately. For example, when calculating the average length of stay in the county jail for inmates charged with a class C felony during a particular year, an analyst discovers that all inmates spent less than three months except for one inmate who stayed for 25 months. This one value made the average length of stay to be twelve months.

Is the value twelve months truly typical of how long inmates were staying in jail?

Answer:


Possible reasons for outliers:

1. The researcher sampled from a "different" population without realizing it. The outlier should not be discarded, but if it can be shown that they sampled from the wrong population, then the values will not be used in the original study.

   For example, in Australia researchers were investigating opossums and realized that every once in awhile, there would be an outlier for the tail, ring finger, or ear measurement. Fortunately they did not throw away these outliers and soon saw a pattern in the outliers. From the outliers, they discovered a whole new species of opossum; one that had a longer tail, shorter finger, and shorter ears than the opossums they were studying.(From Chance magazine)

2. A mistake was made while taking a measurement or entering it into the computer. After certain justification, the value should be either corrected or removed from the data set.

   For example, a common measurement error for newborn babies is their length. This is due to the fact that different tape measures have different "starting points" and the nurse may not realize this. Also, newborn babies are sometimes difficult to "stretch out" to measure. (When the author's oldest daughter Ana was born, her length was measured as 22 inches. However for her first week check-up, she was measured as being 21 inches long.)

3. The researcher sampled from the "correct" population and no error was made. The outlier is a legitimate data value from the population and represents natural variability in the data. In this case, the data value should not be discarded because it provides important information about the population.

   For example, when the author's daughter Ana was born she weighed 11 lbs, 1 oz. She was not diabetic and neither was Ana. Ana was simply just a big baby!

*Outliers should never be discarded without proper justification!*

The Mode:

- May not be unique
- The mode is less affected by outliers than either the mean or median. Unfortunately, the mode will vary more from sample to sample (from the same population) than either the mean or the median, making it a poor choice as a measure of central tendency.

- Measures of Dispersion – conveys information about the amount of variability in the data. There are four different measures of dispersion we will discuss.

  1. Range
  2. Variance
  3. Standard Deviation
  4. Interquartile Range

## The Range

- The range measures the variability by comparing the largest and smallest data value.

  $R = x_L - x_S$ (i.e., the largest data value minus the smallest data value)

  Is the range a resistance measure?

  Answer:

## The Variance

- The variance measures the variability by comparing each data value to the mean. Just like with the mean there are two variances: a population variance and a sample variance.

  $$\sigma^2 = \frac{\sum_{i=1}^{N} (x_i - \mu)^2}{N}$$  $\sigma^2$ is the population variance

  Is $\sigma^2$ a parameter or a statistic?

  Is $\sigma^2$ a constant or a random variable?

  $$s^2 = \frac{\sum_{i=1}^{n}(x_i - \bar{x})^2}{n-1} = \frac{n\left(\sum x_i^2\right) - \left(\sum x_i\right)^2}{n(n-1)}$$  $s^2$ is the sample variance

  Is $s^2$ a parameter or a statistic?

  Is $s^2$ a constant or a random variable?

  Example of finding $s^2$: Suppose the data set is as follows: 1     2     3.

  then $s^2 = \dfrac{3(1^2 + 2^2 + 3^2) - (1 + 2 + 3)^2}{3(2)} = \dfrac{3(14) - 6^2}{6} = \dfrac{42 - 36}{6} = 1$

  - Is the variance a resistance measure?

    Answer:

The Standard Deviation

- The standard deviation measures the variability in the original units of the data

  $\sigma = \sqrt{\sigma^2}$    $\sigma$ is the population standard deviation

  $s = \sqrt{s^2}$    s is the sample standard deviation

  - Is the standard deviation a resistance measure?

    Answer:


- Coefficient of Variation. When wanting to compare the dispersion (or variability) between two sets of data one needs to be careful because:

  1. The data might have different units. For example, one might be comparing hemoglobin levels measured in gm/dL versus body weight measured in pounds.

  2. The data might have extremely different means. For example, one might be comparing weights of babies versus weights of adults.

- To compare the variability of two data sets, you can calculate the Coefficient of Variation:

  $$C.V. = \frac{s}{\bar{x}}(100)$$

  The C.V. is given as a percentage. The data set with the largest C.V. has the most variability.


- <u>Measures of Location</u> – conveys information about the location of a specific data value compared to the other data values.

  1. Percentiles

  2. Quartiles

  3. Standardized Values

Percentiles

- Percentiles: The $m^{th}$ percentile, $P_m$, is the value $x$ such that $m\%$ of the data values are less than $x$ and $(100 - m)\%$ are greater than $x$.

  If you have your IQ tested, would you rather be told you are in the 98th percentile or the 2nd percentile?

  Answer:

Quartiles

- Quartiles: Quartiles are specific percentiles. The Quartiles divide the data into quarters.

$Q_1 = P_{25}$

$Q_2 = P_{50} =$ _____

$Q_3 = P_{75}$

There are many different methods for finding percentiles/quartiles, which will result in different answers. For this class, we will not compute the percentiles/quartiles by hand. We will use JMP to calculate. We will also use JMP to calculate the mean and variance.

---

To obtain the mean, standard deviation, quartiles and the following percentiles: $P_{0.5}, P_{2.5}, P_{10}, P_{97.5}, P_{99.5}$, perform the following steps:

1. Enter the data. Under the Analyze menu, choose Distribution.

2. Click on the column containing the data, click on Y, hit ok.

3. (To obtain the other measures discussed above, under the display option, choose more moments.)

---

To obtain a specific percentile, do the following:

1. Click on new column.

2. Under column Properties, choose formula.

3. Choose statistical and then choose col(Quantile).

4. Click on column containing data and click on first box.

5. In second box, type in the desired percentile as a decimal.

---

Let's illustrate the above steps with the following example:

- Suppose the following data represents the age at which a randomly selected American adult first received a speeding ticket.

| 18 | 18 | 19 | 19 | 20 | 21 | 21 | 21 |
|----|----|----|----|----|----|----|----|
| 22 | 22 | 23 | 23 | 23 | 32 | 32 | 32 |
| 32 | 33 | 36 | 36 | 37 | 39 | 39 | 40 |
| 41 | 42 | 42 | 43 | 44 | 44 | 45 | 46 |
| 48 | 49 | 50 | 51 | 52 | 56 | 56 | 61 |
| 63 | 65 | 88 |    |    |    |    |    |

Use JMP to find the requested percentiles and the quartiles (and moments):

JMP output:

| Distributions age | Moments | Distributions | age | Quantiles |
|---|---|---|---|---|
| Mean | 38.232558 | 100.0% | maximum | 88 |
| Std Dev | 15.502438 | 99.5% | | 88 |
| Std Err Mean | 2.3640996 | 97.5% | | 85.7 |
| Upper 95% Mean | 43.003504 | 90.0% | | 59 |
| Lower 95% Mean | 33.461612 | 75.0% | quartile | 48 |
| N | 43 | 50.0% | median | 39 |
| Sum Wgt | 43 | 25.0% | quartile | 23 |
| Sum | 1644 | 10.0% | | 19.4 |
| Variance | 240.32558 | 2.5% | | 18 |
| Skewness | 0.7913035 | 0.5% | | 18 |
| Kurtosis | 1.0041674 | 0.0% | minimum | 18 |
| CV | 40.547739 | | | |
| N Missing | 0 | | | |

- Find and interpret the 10$^{th}$ percentile, $P_{10}$:

   Answer:

- Find the Quartiles:

   Answer:

Recall, when discussing Measures of Dispersion, we did not define the Interquartile Range. Now, we can.

- *The Interquartile Range*:

$$IQR = Q_3 - Q_1$$

The *IQR* describes the variability of the middle 50% of the data values.

Is the *IQR* a resistance measure?

Answer:


What Measure of Dispersion should we use if our data value contains outliers?

Answer:


If the data set does not contain outliers, what measure of dispersion should we use, the IQR or the variance? Why?

Answer:


# ➡ Section 5: Distribution Shapes:

When analyzing data, it is important to determine the shape of the data, since the shape can affect which measure we choose to describe the data and which method we use to analyze the data.

We determine the shape of the distribution by visualizing the histogram smoothed out into a curve, as shown in the graph below.

- distribution shapes
  - frequency "polygons" or "ogives"

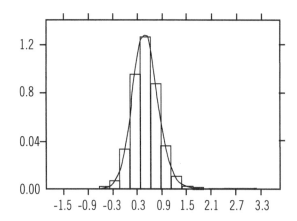

- unimodality and skewness

Symmetric

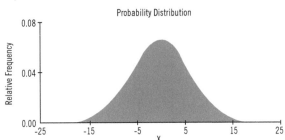

positive (right) skewed

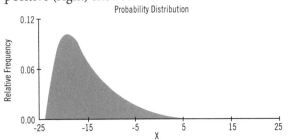

negative (left) skewed

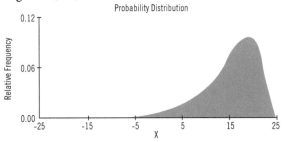

- As we will see later, we like the distribution of our data to be symmetric and not skewed.

- If the data is skewed, we should use the _____ as the measure of central tendency. Why?

  Answer:

- If the data is skewed, we should use the _____ as the measure of dispersion. Why?

  Answer:

- If the data is symmetric, we should use the _____ as the measure of central tendency. Why?

  Answer:

- If the data is symmetric, we should use the _____ as the measure of dispersion. Why?

  Answer:

More shapes:

- multimodal

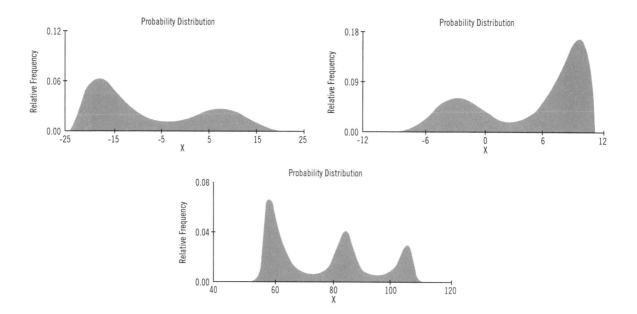

- Typical relationships among mean, median, and mode

    **a.** Positive (right) skewed     <u>usually</u> ↦ mode < median < mean

    **b.** Negative (left) skewed     <u>usually</u> ↦ mean < median < mode

    **c.** Symmetric and Unimodal    <u>usually</u> ↦ mean ≈ median ≈ mode

- Exceptions to the mean, median, mode inequalities

    - bimodal distributions

    - This distribution is approximately symmetric but not unimodal

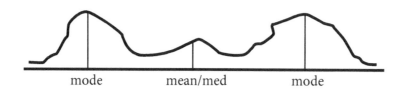

    - It is easy to construct a distribution which violates a or b:

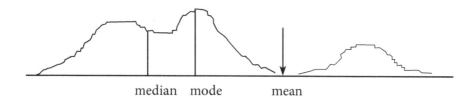

# ➡ Section 6: Modified Boxplots:

A "picture" of the data that displays the shape, the middle, the variability of the data and whether the data contains outliers. To find, we need to calculate the following:

1. The Quartiles — $Q_1$, $Q_2$, and $Q_3$

2. *IQR*

3. $Q_1 - 1.5*IQR$     }     These values are used for detecting outliers. Any value less than $Q_1 - 1.5*IQR$ is a possible outlier. Any value greater than $Q_3 + 1.5*IQR$ is a possible outlier.

4. $Q_3 + 1.5*IQR$

- A boxplot:

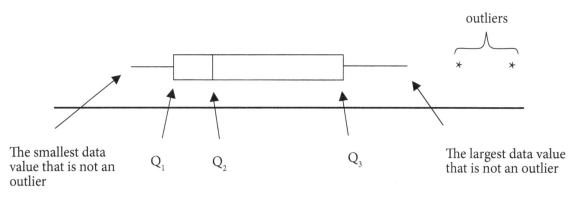

The smallest data value that is not an outlier     $Q_1$     $Q_2$     $Q_3$     The largest data value that is not an outlier

outliers

---

- Steps for using JMP to find the Modified Boxplot:

  1. Enter the data in the column.

  2. Under the graph menu, choose graph builder.

  3. Click on the column containing the data and drag to the x-axis.

  4. At the top of the window, click on the modified boxplot button.

  5. To copy the boxplot, click on the done button. Then you can select and copy.

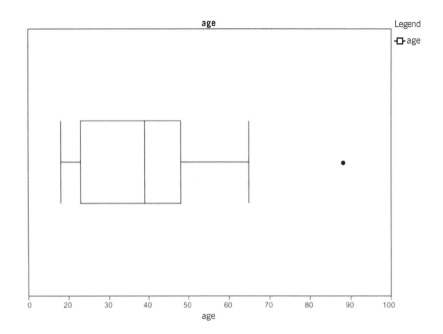

From the modified boxplot, we can describe the "typical" data value and the variability of the data using resistant measures.

We can discuss the quartiles and talk about the shape of the distribution. We can also determine at a glance if there are any outliers.

By asking for the moments, we can specify the above very specifically:

| Distributions age Quantiles | | |
|---|---|---|
| 75.0% | quartile | 48 |
| 50.0% | median | 39 |
| 25.0% | quartile | 23 |

$Q_1 =$

$Q_2 =$

$Q_3 =$

We can calculate the IQR:   IQR =

Interpretation of IQR:

We can determine the bounds for any outliers:

(i) $Q_1 - 1.5(\text{IQR}) =$

(ii) $Q_3 + 1.5(\text{IQR}) =$

- Here are several examples contrasting box plots with underlying shapes

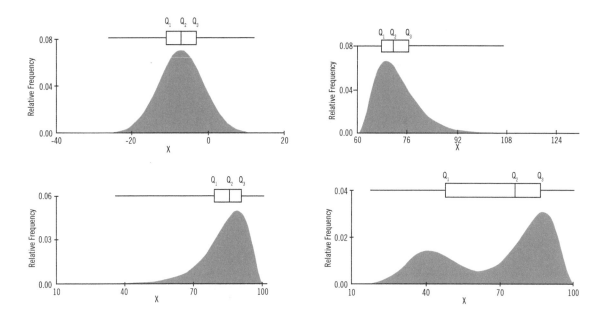

# Section 7: Two Very Important Facts about Describing Data:

- Fact 1: When describing a data set you should always do the following three steps:

  1. Obtain a "picture" of the data

  2. Calculate a measure of central tendency

  3. Calculate a measure of dispersion.

  You should never describe a data set without doing all three steps or your description could be incorrect or at the least, very misleading.

- Fact 2: Be very careful looking at trends in averages of some quantity (like income) over time. What you may be observing is an increase in variability and skewness over time, resulting in an increase in averages, but leaving the modal values essentially unchanged. You would be able to detect that phenomenon if you had the three necessary tools mentioned above.

For example suppose you have the graphs below that represent the waiting times at a doctor's office for three consecutive months. The graphs on the left describe the waiting times for Dr. Allison. The solid vertical line represents the averages. The graphs on the right describe the waiting times for Dr. Barnes. The solid vertical line represents the average and the dashed line represents the mode.

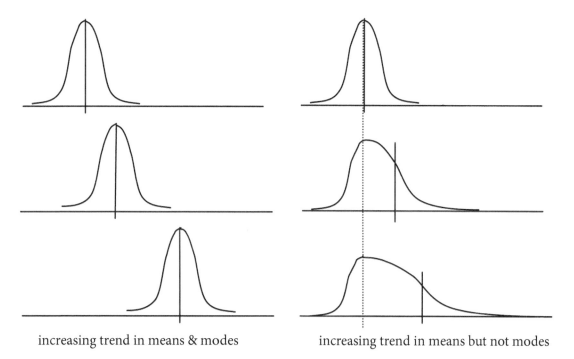

increasing trend in means & modes    increasing trend in means but not modes

If we just look at the solid vertical lines, how do the average waiting times compare for Dr. Allison and Dr. Barnes?

Answer:


What do the graphs on the left "tell us" about the overall waiting times for Dr. Allison?

Answer:


What do the graphs on the right "tell us" about the overall waiting times for Dr. Barnes?

Answer:


Taking into account the variability and the shape, can we conclude the waiting times for the two doctors are the same?

Answer:

# ➡️ Chapter 1 Homework:

1. Dr. Green is concerned about the sleeping habits of STA 1380 students. Altogether, there are five sections of STA 1380, each containing thirty students. Dr. Green decides to randomly select ten students from each of the five different classes. She finds that 38% of the sampled students sleep less than six hours per night on average. Upon hearing this, the head of the department is greatly concerned and checks the sleeping hours of all the STA 1380 students and finds only 13% are getting less than six hours of sleep per night.

_____ **i.** The population in this study is _____.

   **A.** The fifty students surveyed.

   **B.** The ten students selected from Dr. Green's class.

   **C.** The entire student body of Baylor.

   **D.** All students currently enrolled in STA 1380.

_____ **ii.** The sample in this study is _____.

   **A.** The fifty students surveyed.

   **B.** The ten students selected from Dr. Green's class.

   **C.** The entire student body of Baylor.

   **D.** All students currently enrolled in STA 1380.

_____ **iii.** In the paragraph above, Dr. Green's initial finding of 38% is a(n) _____.

   **A.** Error.

   **B.** Statistic which is a random variable.

   **C.** Statistic which is a constant.

   **D.** Parameter which is a random variable.

   **E.** Parameter which is a constant.

   **F.** Random Variable.

_____ **iv.** The head of the department's finding of 13% is a(n)_____.

   **A.** Error.

   **B.** Statistic which is a random variable.

   **C.** Statistic which is a constant.

   **D.** Parameter which is a random variable.

   **E.** Parameter which is a constant.

   **F.** Random Variable.

_____ **v.** What type of sampling scheme did Dr. Green use?

   **A.** Simple Random Sampling.

   **B.** Systematic Sampling.

   **C.** Stratified Sampling.

   **D.** Cluster Sampling.

   **E.** Sample of Convenience.

**2.** Mrs. Strong is concerned about the eating habits of the students at Crawford Elementary. Altogether, there are twenty different classes of kids, each containing thirty students. Mrs. Strong decides to randomly select twelve classes and sample each child in the sampled class. She finds that 48% of the sampled students ate a healthy breakfast before coming to school. Upon hearing this, the superintendent, Dr. Jones, is greatly concerned and checks the eating habits of all the elementary students and finds that 63% ate a healthy breakfast before coming to school.

_____ **i.** The population in this study is _____.

   **A.** The 360 students surveyed.

   **B.** All the elementary students at Crawford Elementary.

   **C.** The 30 students selected from Mrs. Strong's class.

   **D.** The entire student body of Crawford ISD.

   **E.** None of the above.

_____ **ii.** The sample in this study is _____.

   **A.** The 360 students surveyed.

   **B.** All the elementary students at Crawford Elementary.

   **C.** The 30 students selected from Mrs. Strong's class.

   **D.** The entire student body of Crawford ISD.

   **E.** None of the above.

_____ **iii.** In the paragraph above, Mrs. Strong's initial finding of 48% is a(n) _____.

   **A.** Error.

   **B.** Statistic, which is a constant.

   **C.** Statistic, which is a random variable.

   **D.** Parameter, which is a constant.

   **E.** Parameter, which is a random variable.

   **F.** None of the above.

_____ **iv.** Dr. Jones's finding of 63% is a(n) _____.

    **A.** Error.

    **B.** Statistic, which is a constant.

    **C.** Statistic, which is a random variable.

    **D.** Parameter, which is a constant.

    **E.** Parameter, which is a random variable.

    **F.** None of the above.

_____ **v.** What type of sampling scheme did Mrs. Strong use?

    **A.** Simple Random Sampling.

    **B.** Systematic Sampling.

    **C.** Stratified Sampling.

    **D.** Cluster Sampling.

    **E.** Sample of Convenience.

**3.** Fill in the blank with the correct symbol.

    **a.** We denote the population mean with the symbol _____.

    **b.** We denote the sample mean with the symbol _____.

    **c.** We denote the population variance with the symbol _____.

    **d.** We denote the sample variance with the symbol _____.

    **e.** We denote the population size with the symbol _____.

    **f.** We denote the sample size with the symbol _____.

    **g.** We denote the population median with the symbol _____.

    **h.** We denote the sample median with the symbol _____.

    **i.** We denote the population standard deviation with the symbol _____.

    **j.** We denote the sample standard deviation with the symbol _____.

**4.** Identify the correct term for the type of bias/effect being defined.

    **a.** A survey is mailed to 1000 Waco residents. Only 248 of the surveys are returned.

    **b.** A secondary researcher noticed that subjects changed their answers when the lead researcher raised his eyebrows upon hearing their first response.

    **c.** The store manager at Big Lots randomly surveyed twenty-five morning shoppers to determine how all of the shoppers felt about the new store hours.

    **d.** Mr. Hall surveyed fifty high school students asking them if they have ever skipped school.

    **e.** Students who were viewed cheating by camera did not cheat when people were brought in to monitor their exams.

    **f.** Although the child only received a sugar pill, she claimed she no longer had a headache.

**5.** For the following scenarios, choose which measure should be calculated.

    **a.** The goal is to describe the "typical" value in the data set and the data set contains an outlier.

    **b.** The goal is to describe the amount of variability in the data set and the data set is symmetric.

    **c.** The goal is to describe the true center, the balancing point of the data and the data set is symmetric.

    **d.** The goal is to describe the amount of variability in the data set and the data set is skewed right.

**6.** For the following situations, choose the sampling scheme or method used to obtain data.

    **a.** To understand the social dynamics of monkeys, a researcher observed a tribe of monkeys in Africa for five days.

    **b.** From a group of 200 senior adults living at the nursing home, every tenth adult starting with the fourth adult is surveyed to determine their opinion on the new social director.

    **c.** To poll the nurses working at Scott and White, a researcher used a random number generator to randomly selected one hundred nurses.

    **d.** A large corporation with 1000 employees divided the employees into ten groups based on the type of work done and sent a questionnaire to twenty randomly selected employees from each of the ten groups to determine what all the employees' opinions are on the new insurance requirements.

    **e.** To determine what his students feel about his new grading policy, Dr. Smith surveyed each of his students in his class.

    **f.** To understand what their customers want, Grey Hound Bus Line, randomly selected fifty buses and surveyed every passenger on the bus.

    **g.** To determine what Baylor students feel about the new proposed mascot emblem used on all official athletic wear, Dr. Smith samples the students in his classes.

**7.** Suppose the following data set gives the age a person first receives a winning lottery ticket. Use JMP to answer the questions below. Make sure and print your output!

| 35 | 43 | 16 | 30 | 20 | 32 | 25 | 26 | 32 | 30 | 18 | 35 |
|----|----|----|----|----|----|----|----|----|----|----|----|
| 21 | 27 | 34 | 5  | 38 | 31 | 38 | 41 | 15 | 22 | 28 | 38 |

    **a.** Find $P_{15}$.

    **b.** In the context of this problem, interpret $P_{15}$.

    **c.** Find the Quartiles.

    **d.** Find the Interquartile Range.

    **e.** Interpret the Interquartile Range.

    **f.** Determine if the data set contains any outliers. Show your work. If there are outliers, make sure and state which data value(s) are outliers.

    **g.** Construct a Boxplot (Modified Boxplot).

    **h.** Find the typical data value of the data set.

**8.** On the last day of class, two graduate statistics classes at the University of Texas at Arlington were given a survey. One class was made up of statistic majors, while the other was made up of non-statistic majors. One of the questions asked was, "During the semester, how many hours did you study per week, to the nearest hour?" Use the modified boxplots below to answer the questions that follow.

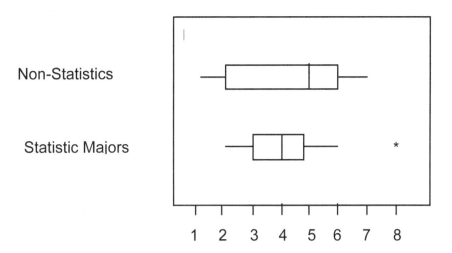

**a.** Estimate the median hours studied for non-statistic major students.

**b.** Which group has more variability in the middle 50% of the data?

**c.** Approximately what percent of non-statistic majors studied less than all the statistic majors sampled?

**d.** Is there an outlier present in either the statistic majors or the non-statistic majors? If so, what is the value of the outlier?

**e.** What is the approximate shape of the distribution for the statistic majors?

**f.** What is the number of hours such that only 25% of the non-statistic majors studied more than that per week?

**9.** Suppose data set 1 represents the amount of weight lost in two weeks for twenty randomly selected people following a low carbohydrate diet. Suppose data set 2 represents the amount of weight lost in two weeks for twenty selected people following a low fat diet. Use JMP to answer the following questions.

| Data Set 1: | 7 | 9 | 6 | 6 | 5 | 4 | 7 | 6 | 7 | 5 |
|---|---|---|---|---|---|---|---|---|---|---|
|  | 6 | 6 | 5 | 8 | 4 | 6 | 7 | 6 | 7 | 7 |

| Data Set 2: | 3 | 4 | 2 | 5 | 4 | 5 | 4 | 3 | 4 | 3 |
|---|---|---|---|---|---|---|---|---|---|---|
|  | 5 | 3 | 3 | 4 | 6 | 3 | 4 | 7 | 6 | 3 |

**a.** For each data set, find the average amount of weight lost. Is the value calculated $\mu$ or $\bar{x}$?

**b.** For each data set, find the variability of the data. Is the value calculated $\sigma^2$ or $s^2$?

**c.** For each data set, find the coefficient of variation. Which data set has more variability?

**d.** Use JMP to create a modified boxplot for each data set on the same axis.

**e.** Using the modified boxplots, compare and contrast the two diets. Which diet appears to work better? Justify your answers.

10. For the following histograms, determine the following:

   i. The shape of the distribution.

   ii. Whether $\bar{x} < \tilde{x}$ or $\bar{x} > \tilde{x}$ or $\bar{x} \approx \tilde{x}$.

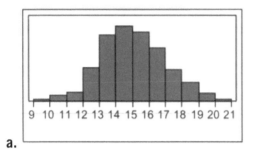

a.

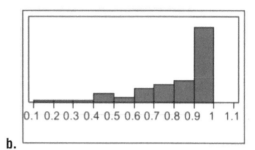

b.

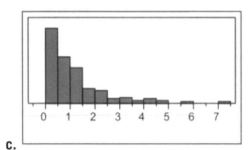

c.

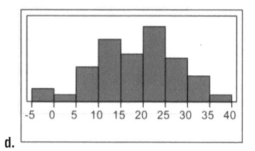

d.

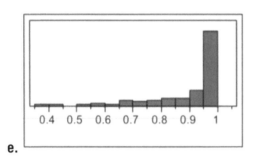

e.

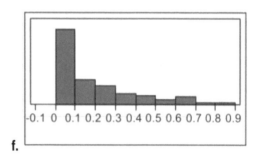

f.

11. _____ If the goal is to describe how a particular data value is related to the other data values in the data set, which measure should be calculated?

   a. Measure of Central Tendency.

   b. Measure of Dispersion.

   c. Measure of Location.

12. _____ If the goal is to describe the typical value in a data set, which measure should be calculated?

   a. Measure of Central Tendency.

   b. Measure of Dispersion.

   c. Measure of Location.

**13.** _____ If the goal is to describe the amount of variability in a data set, which measure should be calculated?

    **a.** Measure of Central Tendency.

    **b.** Measure of Dispersion.

    **c.** Measure of Location.

**14.** Fill in the blank with either the word _parameter_ or _statistic_:

    **a.** The _____ is a numerical value based on the sample.

    **b.** The _____ is a numerical value based on the population.

    **c.** In the field of Statistics, we use the _____ to draw conclusions about the _____.

    **d.** $\mu$ is a _____.

    **e.** $\bar{x}$ is a _____.

    **f.** $\sigma$ is a _____.

    **g.** $s$ is a _____.

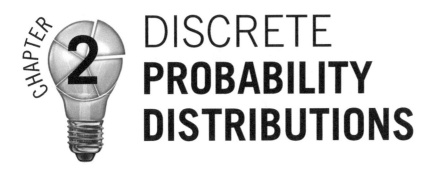

# DISCRETE PROBABILITY DISTRIBUTIONS

CHAPTER 2

## ➡ Section 1: Introduction to Probability:

The theory of probability provides the foundation for statistical inference. I like to think of it as a tool in our tool belt we use for drawing conclusions.

Whenever we draw a conclusion in statistics, we have no idea if we have made the correct decision or an incorrect decision and thus we quantify that uncertainty about our decision using probability. For our purposes, we will only consider the fundamental aspects of probability.

Consider the following: Suppose six out of ten patients suffering from some disease are cured after receiving a particular treatment. Could this have happened even without the treatment or is this evidence that the treatment is effective?

We answer these types of questions using the concepts and laws of probability.

But first, what exactly is probability?

Most statisticians define three different types of probability:

1. *Subjective* probability, which is a measure of personal uncertainty.

2. Probability as *frequency*, which is based on the number of times that a particular event happens during repeated trials of some experiment. This is the type of probability we will use because it is the most widely used in actual practice.

3. *Classical* probability, which uses counting techniques for calculating probability. This is historically associated with games of chance.

## ➡ Definitions:

- *Probability*: a measure of uncertainty.
  - Probability: is always between 0 and 1, inclusive.
  - If $P(X = a) = 0$, then that implies that "*a*" _____ occur.
  - If $P(X = a) = 1$, then that implies that "*a*" and only "*a*" _____ occur.
  - If $P(X = a)$ is "large," then that implies that "*a*" is   very likely or   not very likely   to occur.
  - If $P(X = a)$ is "small," then that implies that "*a*" is   very likely or   not very likely   to occur.
- *Experiment*: a process in which the outcome can not be predicted ahead of time.

- *Sample Space:* $S$ = a collection of all possible outcomes.
  - If describing $S$, are you describing a population or a sample?
    Answer:


  - $N$ = the size of the sample space.
- *Event:* $E$ = a subset of the sample space.
- *Equally Likely Events:* events are equally likely when each outcome in the event has the same chance of occurring.
- *Mutually Exclusive Events:* events that cannot happen at the same time. In other words, there are no elements in common between the events.
- *Exhaustive Events:* when every element in $S$ occurs in one of the events. In other words, there are no elements in $S$ that are not in one of the events listed.
- For example: Suppose you roll a die. Let $E_1$ = the die is odd and let $E_2$ = the die is even. Then

  $E_1 = \{1, 3, 5\}$ and $E_2 = \{2, 4, 6\}$

  $E_1$ and $E_2$ are equally likely events that are also mutually exclusive and exhaustive.

# ➡ Elementary Properties of Probability:

1. (*The Convexity Law*) Given some process or experiment with $n$ mutually exclusive outcomes (called events), $E_1, E_2, ..., E_n$,

   $0 \le P(E_i) \le 1$

2. Given some process or experiment with $n$ mutually exclusive outcomes,

   $P(E_1) + P(E_2) + ... + P(E_n) = 1$

3. For any two events, A and B

   $P(A \ or \ B) = P(A \cup B) = P(A) + P(B) - P(A \cap B)$

4. (*The Multiplication Rule*) For any two events, $A$ and $B$

   $P(A \ and \ B) = P(A \cap B) = P(A)P(B \mid A) = P(B)P(A \mid B)$

5. (*The Multiplication Rule for independent events*)

   $P(A \ and \ B) = P(A \cap B) = P(A) \cdot P(B)$

6. *Probability of Equally Likely Events:* $P(E) = \dfrac{f}{N}$.

- Examples:

                         <u>Equally likely Events?</u>                <u>Probability?</u>

- flipping a coin

  $S =$

- rolling a die

  $S =$

- shooting a free throw shot

  $S =$

# ➡ Section 2: Probability Distributions:

Suppose we flip a coin five times. What is $S$?

Answer:

As we can see, listing all the values of $S$ can be daunting. Fortunately, it is not always necessary. Recall from chapter one that a random variable is a variable that is obtained as a result of a chance process. When calculating probabilities, instead of listing $S$ to describe the population, we will use random variable notation to describe the population. Here is another definition of random variable:

*Random variable* – a rule that assigns a numerical quantity to each outcome in the sample space. We generally denote random variables with $X$, $Y$, or $Z$.

- Example: Suppose we flip a coin three times and let $X$ = the number of heads.

  $S =$

  $X =$

- Random variables are either discrete or continuous.
- In inferential statistics, we are trying to draw conclusions about the population based on what we have seen from our sample.
- We represent our uncertainty about our conclusions with probability. Thus, to every value of a variable, we can attach a probability representing our uncertainty about that value.

  This is the *probability distribution* of the random variable.

- The *probability distribution* describes the population of data values. In other words, the probability distribution describes *all* the possible values that can be attained by the random variable.

  - Is the probability distribution describing the population or a sample?

    Answer:

The probability distribution depends on the type of random variable obtained:

- If the random variable is continuous, we describe its distribution with a continuous curve, called a *probability density function* (PDF).

- If the random variable is discrete, we describe its distribution with a probability mass function (PMF).

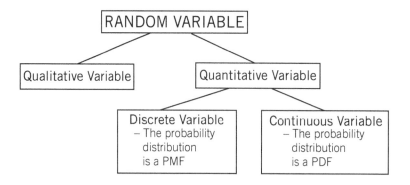

# ➡️ Section 3: Discrete Probability Distributions:

<u>Probability Mass Function</u> – The probability distribution of a discrete random variable, called a *Probability Mass Function*, is a table, graph, formula, or any device used to specify all possible values of a discrete random variable along with their respective probabilities.

**Notation:** $f(x) = P(X = x)$

For Example:

| $X$ | 2 | 3 | 4 | 5 | 6 |
|------|------|------|------|------|------|
| $f(x)$ | 0.1 | 0.4 | 0.1 | 0.3 | 0.1 |

$f(3) =$

The graph of a PMF is called a *Probability Histogram.*

To make a probability histogram:

1. The horizontal axis represents the random variable.

2. The vertical axis represents the probability.

**3.** The probability is represented by a bar with width one that is centered about the random variable and has height equal to the probability.

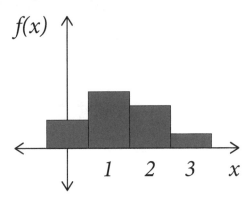

- Example 1: Consider flipping a coin three times. Let $X$ = # of heads.

  - Find the PMF:

    Answer:

  - Graph the PMF:

    Answer:

- Properties: For a function to be a PMF, the following must be true.

  **1.** $0 \leq f(x) \leq 1$

  **2.** $f(x) = 0$ for all $x$ not in $S$

  **3.** $\sum_{x \in S} f(x) = 1$

- Example 2: Ava has a four sided green die with the numbers 1, 2, 3, 4 on the sides. (One side has a 1, one side has a 2, etc.) Ali has a six sided pink die with the numbers 1, 2, 3, 4, 5, 6. Suppose Ava and Ali each roll their die one time. Let $X$ = sum of the green and pink dice. Find the PMF for $X$.

  Answer:

- Cumulative Distribution Function: *CDF* – the cumulative distribution function of a discrete random variable is a table, graph, formula, or any device used to specify all possible values of a discrete random variable along with their respective cumulative probabilities.

    $F(x) = P(X \leq x) = \Sigma f(x)$

    For example:

| X | 2 | 3 | 4 | 5 | 6 |
|------|-----|-----|-----|-----|-----|
| F(x) | 0.1 | 0.5 | 0.6 | 0.9 | 1.0 |

    $F(3) =$

- Example 3: Find *F(x)* for our example 1:

    Answer:

- The Graph of a CDF is a step function:

    Answer:

- Properties of *F(x)*, where *F(x)* is a CDF:

    **1.** $0 \leq F(x) \leq 1$
    **2.** $\lim_{x \to -\infty} F(x) = 0$ and $\lim_{x \to +\infty} F(x) = 1$
    **3.** If $a < b$, $F(a) \leq F(b)$
    **4.** *F(x)* is right continuous

- Example 4: Find $F(x)$ for our example 2:

  Answer:

- Comparing f(x) versus F(x):

  Recall, $f(x) =$

  Whereas

  $F(x) =$

- Example 5: Finding the probability with $f(x)$ versus $F(x)$: Suppose you are given the following PMF for the RV $X$, where $X$ represents the number of successes.

  a. Find the CDF

  | X | f(x) |
  |---|------|
  | 0 | 0.2  |
  | 1 | 0.1  |
  | 2 | 0.3  |
  | 3 | 0.1  |
  | 4 | 0.1  |
  | 5 | 0.2  |

  b. Find $P(2 < X < 5)$ using $f(x)$ and $F(x)$.

  Answer:

  c. Find probability of having at least 3 successes:

  Answer:

**d.** Find probability of having at most 2 successes:

Answer:

**e.** Find probability of having between 2 and 4 success, inclusive.

Answer:

**f.** Find probability of having between 1 and 4 success, exclusive.

Answer:

Mean and Variance of a Discrete Probability Distribution:

- When describing a sample, one should always include three things: a "picture" of the data using a graph, a measure of central tendency, and a measure of dispersion. This is true when describing a population. The probability histogram gives us a "picture" of the data (or the step graph of the CDF). We also would like to know a measure of central tendency and a measure of dispersion. The measure of central tendency we will discuss is the population mean, which is also known as the expected value. The measure of dispersion we will discuss is the variance or standard deviation.

The Expected Value: It is often of interest to find the long-run average of a particular distribution.

- This long-run average is known as the expected value or mean value of the random variable
  - Formally, the expected value of a random variable $X$ is the mean value of the variable in the sample space or population of possible outcomes. Think of it as the mean of what we would get if we did the experiment over and over and over again...what we would get in the long run...what we *expect* to get in the long run.
- Calculating Expected Value

  If $X$ is a random variable with possible values $x_1$, $x_2$, $x_3$,... occurring with probabilities $f(x_1), f(x_2), f(x_3)$, ...then

$$E(X)=\sum x_i f(x_i)$$

- Variance and Standard Deviation for a Discrete Random Variable

  - Recall that variance and standard deviation describe how much variability there is in the data by comparing how spread out the values are about the mean. This gives us an idea of how far from the mean we expect values to fall for data from a given distribution.
  - Calculating variance: If $X$ is a random variable with possible values $x_1$, $x_2$, $x_3$,... occurring with probabilities $f(x_1), f(x_2), f(x_3)$, ...and with expected value $E(X) = \mu$ then

$$\text{Variance of } X = V(X)= \sigma^2=\sum(x_i-\mu)^2 f(x_i) = \sum x^2 f(x)-\mu^2$$

  - Standard deviation of $X$: $\sigma = \sqrt{\sum x^2 f(x)-\mu^2}$
- Example 6: Suppose you are going to gamble on a roulette wheel, where you pay $1.00 to place your bet. (The roulette wheel has thirty-eight equal spaces, eighteen red, eighteen black, and two green.) If the ball falls on the black eight, you win $3.00. Let $X =$ the *net* winnings.

Find and interpret the expected value:

Find the variance:

# Check Your Understanding

1. Suppose the cumulative distribution function for the random variable $X$ is as follows:

| $X$ | 0 | 0.5 | 1 | 1.5 | 2 | 4 |
|-----|-----|------|------|------|------|------|
| $F(x)$ | 0.47 | 0.62 | 0.73 | a | 0.98 | b |

   a. If the $P(X = 1.5) = 0.08$ what is the value of a?

   b. What is the value of b?

   c. Find $P(X = 2.5)$

   d. Find the probability that $X$ is at least 1.5.

   e. Find $P(0 < X \leq 2)$

2. Suppose you play a dice game where if you roll a composite number, you win $5.00. If you roll a one, you win $10. If you roll a prime number, you lose $8. You must pay $2.00 to play. Let $X$ = net winnings.

   a. Find the PMF.

   b. Find the expected value.

   c. Interpret the expected value.

## ➡ Section 4: The Binomial Distribution:

When finding the probability mass function, finding the probability can sometimes be challenging. Some experiments have specific characteristics. These specific characteristics can be modeled by a specific, well-defined discrete probability distribution. The benefit to this is that if we recognize the characteristics of a well defined distribution, then there will be a known formula or table that we can use for calculating the probabilities. One such well known discrete probability distribution is the Binomial distribution.

- *Bernoulli Trial* – when a random process or experiment results in one of only two mutually exclusive outcomes, the trial is called a *Bernoulli Trial.*

- *Bernoulli Process* – a sequence of Bernoulli trials such that:

   1. Each trial has only two mutually exclusive outcomes. One is noted as a success, the other as a failure.

   2. The probability of success, $p$, remains constant from trial to trial. The probability of failure is denoted as $q$.

   3. Trials are independent. The outcome of one trial is not affected by the outcome of any other trial.

- *Binomial Distribution* – if our random variable, $X$, is defined to count the number of successes and our experiment is a finite Bernoulli process, then the resulting distribution is a *Binomial distribution.*

- If the random variable $X$ is a Binomial random variable, we denote it as $X \sim Bin(n, p)$.

Example 1: Are the following random variables binomial random variables?

    **a.** Sixty percent of all commercial bus drivers are over thirty years of age. Out of a random sample of twenty-five drivers, let the random variable $X$ be the number of bus drivers who are over thirty years of age.

    Answer:

    **1.** Is the RV counting the number of success?

    **2.** Is it a finite Bernoulli Process?

      **i.** Finite number of trials?

      **ii.** Only two outcomes?

      **iii.** Constant probability?

      **iv.** Independent trials?

    **b.** Suppose a clothing salesperson makes sales to 35% of her customers. One day she counted her customers until she made three sales. Let $X$ be the number of customers until her third sale.

    Answer:

    **1.** Is the RV counting the number of success?

    **2.** Is it a finite Bernoulli Process?

      **i.** Finite number of trials?

      **ii.** Only two outcomes?

     **iii.** Constant probability?

     **iv.** Independent trials?

**c.** A department of a large corporation has ten employees, seven women and three men. Four employees are to be selected to form a committee. Let the random variable $X$ be the number of women on the committee.

Answer:

**1.** Is the RV counting the number of success?

**2.** Is it a finite Bernoulli Process?

     **i.** Finite number of trials?

     **ii.** Only two outcomes?

     **iii.** Constant probability?

     **iv.** Independent trials?

- The Binomial Probability Mass Function:
  The PMF for the Binomial distribution is

$$P(X = x) = f(x) = \binom{n}{x} p^x q^{n-x} \quad X = 0, 1, ..., n$$

where $n$ = the number of trials

     $p$ = the probability of success

     $q$ = the probability of failure.

- If we know $X$ is a Binomial random variable, then we know the summary parameters as well:

    The mean: $\qquad\qquad\qquad\qquad\quad \mu = n{\cdot}p$

    The variance: $\qquad\qquad\qquad\qquad \sigma^2 = n{\cdot}p{\cdot}q$

    The standard deviation: $\qquad\qquad \sigma = \sqrt{n{\cdot}p{\cdot}q}$

---

- We will use JMP to find the probabilities. We can find the probability using the PMF or the CDF. Here are the steps for finding the PMF and CDF values:

    - create a column representing the values of $X$.

    - under *columns*, choose *new column*, name PMF or CDF

    - under *properties*, choose *formula*

    - on the left side of menu, choose *discrete probability* and then choose *binomial probability* for the PMF or *binomial distribution* for the CDF

    - fill in the values for $n$ and $p$

    - for $k$, click on the column representing $X$ and click on $k$.

---

- Example 2: Suppose $X \sim Bin(10, 0.6)$

    JMP Output:

| X | PMF | CDF |
|---|---|---|
| 0 | 0.0001048576 | 0.0001048576 |
| 1 | 0.001572864 | 0.0016777216 |
| 2 | 0.010616832 | 0.0122945536 |
| 3 | 0.042467328 | 0.0547618816 |
| 4 | 0.111476736 | 0.1662386176 |
| 5 | 0.2006581248 | 0.3668967424 |
| 6 | 0.250822656 | 0.6177193984 |
| 7 | 0.214990848 | 0.8327102464 |
| 8 | 0.120932352 | 0.9536425984 |
| 9 | 0.040310784 | 0.9939533824 |
| 10 | 0.0060466176 | 1 |

- Find $P(X = 2)$, using the PMF:

    Answer:

- Find P($X = 2$), using the CDF:

  Answer:

- Find P($X \geq 2$) using the PMF:

  Answer:

- Find P($X \geq 2$) using the CDF:

  Answer:

- Example 3: Seventy percent of the teachers in a large town are opposed to a proposed educational amendment. If twenty teachers are selected at random, find the probability that:

  **a.** More than eighteen teachers are opposed.

**b.** Fewer than two teachers are opposed.

**c.** Between ten and fifteen teachers are opposed.

JMP output:

| X | PMF | CDF |
|---|-----|-----|
| 0 | 3.486784e-11 | 3.486784e-11 |
| 1 | 1.6271661e-9 | 1.6620339e-9 |
| 2 | 3.6068848e-8 | 3.7730881e-8 |
| 3 | 5.0496387e-7 | 5.4269475e-7 |
| 4 | 5.0075583e-6 | 5.5502531e-6 |
| 5 | 0.0000373898 | 0.00004294 |
| 6 | 0.000218107 | 0.000261047 |
| 7 | 0.0010178326 | 0.0012788796 |
| 8 | 0.0038592819 | 0.0051381615 |
| 9 | 0.0120066549 | 0.0171448164 |
| 10 | 0.0308170809 | 0.0479618973 |
| 11 | 0.0653695655 | 0.1133314629 |
| 12 | 0.1143967397 | 0.2277282026 |
| 13 | 0.1642619852 | 0.3919901878 |
| 14 | 0.1916389828 | 0.5836291706 |
| 15 | 0.1788630506 | 0.7624922211 |
| 16 | 0.1304209744 | 0.8929131955 |
| 17 | 0.0716036722 | 0.9645168677 |
| 18 | 0.0278458725 | 0.9923627402 |
| 19 | 0.0068393371 | 0.9992020773 |
| 20 | 0.0007979227 | 1 |

# Check Your Understanding

Montana Table Company produces custom-made furniture for restaurants, diners and hotels. All of their furniture is handmade with it being known that 15 percent of the tables have some sort of flaw. Smitty's Steak House, a large restaurant, orders 15 new tables from Montana Table Company.

    **a.** What is the probability that at least three will be flawed?

    **b.** How many tables ordered by Smitty's Steak House would you expect to have a flaw?

# ➡ Section 5: The Poisson Distribution:

- The *Poisson distribution* describes the number of events that will occur in a specified period of time or in a specified area or in a specified volume.

- Examples of random variables for which the Poisson probability distribution provides a good model are as follows:

    **1.** The number of calls received by the school secretary per day.

    **2.** The number of oak trees with oak wilt per acre.

    **3.** The number of bacteria per one ounce of fluid.

- Characteristics of a Poisson random variable:

    **1.** The experiment consists of counting the number of times a certain event occurs during a given unit of time or in a given area or volume.

    **2.** The probability that an event occurs in a given unit of time, area, or volume is the same for all the units.

    **3.** The number of events that occur in one unit of time, area, or volume is independent of the number of events that occur in other units.

    **4.** The mean or expected number of events in each unit is known and is denoted by $\lambda$.

- *Poisson Distribution* – if the random variable, $X$, is defined to count the number of occurrences of some event per some unit of measurement where the average number of events per unit is known, then the random variable $X$ is said to have a Poisson distribution.

- If the random variable $X$ is a Poisson random variable, we denote it as $X \sim Pois(\lambda)$.

- The Poisson Probability Mass Function:

The *PMF* for the Poisson distribution is

$$P(X = x) = f(x) = \frac{e^{-x}\lambda^x}{x!} \quad X = 0, 1, 2, \ldots$$

where $\lambda$ = the average number of events per unit of measurement (i.e., in each unit of time, area, or volume.)

- If we know $X$ is a Poisson random variable, then we know the parameters:

  The mean: $\mu = \lambda$

  The variance: $\sigma^2 = \lambda$

  The standard deviation: $\sigma = \sqrt{\lambda}$

- The steps for using JMP to find the PMF and CDF for the Poisson Distribution are the same as for the Binomial Distribution. Simply choose *Poisson probability* for the PMF or *Poisson distribution* for the CDF

- Example 1: Suppose the average number of fatal accidents on Section 4 of Interstate 35 is four per month, where the probability of having a fatality per month is independent with constant probability. Find the probability of

    **a.** having no fatalities on Section 4 of Interstate 35 during a one month period.

    **b.** having more than four fatalities on Section 4 of Interstate 35 during a one month period.

**c.** If the researcher is interested in modeling the number of fatalities per two month period,

    **i.** What is the new RV?

    **ii.** What is the distribution of the RV?

JMP Output:

| X | PMF | CDF |
|---|---|---|
| 0 | 0.0183156389 | 0.0183156389 |
| 1 | 0.0732625556 | 0.0915781944 |
| 2 | 0.1465251111 | 0.2381033056 |
| 3 | 0.1953668148 | 0.4334701204 |
| 4 | 0.1953668148 | 0.6288369352 |
| 5 | 0.1562934519 | 0.785130387 |
| 6 | 0.1041956346 | 0.8893260216 |
| 7 | 0.0595403626 | 0.9488663842 |
| 8 | 0.0297701813 | 0.9786365655 |
| 9 | 0.0132311917 | 0.9918677572 |
| 10 | 0.0052924767 | 0.9971602339 |
| 11 | 0.001924537 | 0.9990847709 |
| 12 | 0.0006415123 | 0.9997262832 |
| 13 | 0.0001973884 | 0.9999236716 |
| 14 | 0.0000563967 | 0.9999800683 |
| 15 | 0.0000150391 | 0.9999951074 |
| 16 | 3.7597792e-6 | 0.9999988672 |
| 17 | 8.8465393e-7 | 0.9999997518 |
| 18 | 1.9658976e-7 | 0.9999999484 |
| 19 | 4.1387318e-8 | 0.9999999898 |
| 20 | 8.2774636e-9 | 0.9999999981 |
| 21 | 1.5766597e-9 | 0.9999999997 |
| 22 | 2.866654e-10 | 0.9999999999 |
| 23 | 4.985485e-11 | 1 |

# Check Your Understanding

Suppose the average number of volleyball games in the state of Texas that use the maximum number of officiants is 4.6 per month. Suppose the probability of using the maximum number of officiants per month is constant and independent from month to month.

    **a.** Find the probability that at least 4 games in one month use the maximum number of officiants.

    **b.** Find the probability that at least 8 games use the maximum number of officiants during a three month season.

## ➡ Section 6: Discrete Probability Distributions Overview:

Probability Distributions—    Describes _____ possible values that can be obtained by the random variable.

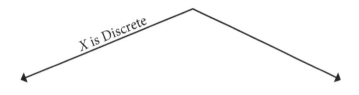

X is Discrete

PMF: $f(x) =$

CDF: $F(x) =$

Suppose $X$ is a discrete RV with possible values, $X = 0, 1, 2, ..., 10$.

Find the following probabilities using $f(x)$ and $F(x)$.

$P(X \leq 2)$:

$P(X > 2)$:

$P(2 < X < 5)$:

Binomial: $X \sim \text{Bin}(n,p)$:

Poisson: $X \sim \text{Pois}(\lambda)$:

## Chapter 2 Homework:

## Section 3

1. Fill in the blank with the correct notation/symbol:

   **a.** We denote the PMF using the notation,

   _____, where _____ = $P(X$ ___ $x)$.

   **b.** We denote the CDF using the notation,

   _____, where _____ = $P(X$ ___ $x)$.

2. Let $f(x)$ given below represent the PMF of the random variable $X$.

$$f(x) = \begin{cases} 1/5 & \text{if } X = -2, -1, 0, 1, 2 \\ 0 & \text{elsewhere} \end{cases}$$

   **a.** Re-write the above PMF as a table.

   **b.** Graph $f(x)$.

   **c.** Find the cumulative distribution function, $F(x)$ of $X$.

   **d.** Graph $F(x)$.

3. Let $f(x) = \dfrac{5!}{x!(5-x)!} \left(\dfrac{1}{3}\right)^x \left(\dfrac{2}{3}\right)^{5-x}$     $X = 0, 1, 2, 3, 4, 5$

   Verify that $f(x)$ is a probability mass function (PMF). Hint: there are two things to verify.

4. Suppose Ana has a pair of dice (the traditional six-sided kind). Let $X =$ the difference of the largest minus the smallest number showing on the dice. Find the PMF for $X$.

**5.** Suppose the probability distribution for $X$ = number of jobs held during the past year for work study students at a university is as follows:

| X | 0 | 1 | 2 | 3 | 4 | 5 |
|---|---|---|---|---|---|---|
| f(x) | 0.12 | a | 0.28 | 0.18 | 0.06 | 0.02 |

    **a.** Find $a$.

    **b.** Find the probability that a randomly selected student held three or more jobs during the past year.

    **c.** Find the cumulative distribution function.

    **d.** Find the $E(X)$.

    **e.** How many jobs would you expect a randomly selected student to hold in the past year?

**6.** A college graduate statistics class has eighteen students. The ages of these students are as follows: one student is 18 years old, two are 20, nine are 22, three are 24, two are 29, and one is 40. Let $X$ = age of any student (randomly selected). Find the probability mass function for $X$.

**7.** Five thousand instant lottery tickets were sold. One ticket has a face value of $10,000, five tickets have face values of $1000 each, twenty tickets are worth $500 each, thirty tickets have face value of $100 each, and 500 are worth $1.00 each. The rest are losers. Let $X$ = the value of a ticket that you buy. Find the probability mass function for $X$.

**8.** Since Carl travels quite a bit for his work, he rents his car. So he is debating whether or not he should purchase car rental insurance. The car rental insurance is $40 a year. He thinks that the probability that he will need to make only one claim a year on his rental car is 0.1, the probability that he has to make only two claims per year is 0.05, and the probability that he will have to make more than two claims a year is zero. It will cost Carl $120 each time if his rental car has to be "fixed" if he doesn't have insurance but will cost nothing if he does have insurance. Let $X$ = Carl's cost next year for "fixing" his rental car and/or car rental insurance.

    **a.** If Carl buys the car rental insurance, what is the PMF? (Hint: $X$ has only one possible value.)

    **b.** If Carl doesn't buy the insurance, what is the PMF? (Hint: $X$ has three possible values.)

    **c.** For parts a and b, find the $E(X)$.

    **d.** Using these two values found in part c, should Carl buy the insurance? Explain why or why not.

**9.** Suppose in a game of cards, if an Ace card is drawn, the player will win $5.00. The player must pay $2.00 to play. Suppose the game is played one time. Let $X$ = the players net winnings.

   **a.** Find the PMF.

   **b.** Find and interpret the E(X).

   **c.** Find V(X).

# Section 4

**10.** Suppose Amy works in the advertisement sales office of the local cable station. Amy successfully sells an advertisement to 65% of her contacts. Suppose Amy has enough time to contact twenty people per day. Let $X$ = the length of time between successful sales of advertisement. Is the random variable $X$ a binomial random variable? If so, give the parameters. If not, explain why not.

**11.** Flying Express Airlines flight 82 from Dallas to Miami is on time 85% of the time, according to Flying Express Airlines. Suppose sixty-five flights are randomly selected.

   **a.** What does the random variable $X$ represent?

   **b.** What is the distribution of the random variable $X$?

   **c.** How many flights from Dallas to Miami would you expect to arrive on time?

**12.** Suppose it is known that a person has a 5% chance of winning a free ticket in a state lottery. Suppose she plays the game fifteen times.

   **a.** What does the random variable $X$ represent?

   **b.** What is the distribution of the random variable $X$?

   **c.** Use JMP to find the PMF.

   **d.** Use JMP to find the CDF.

   **e.** Use the PMF to find the probability that the player wins at most five times.

   **f.** Use the CDF to find the probability that the player wins at least five times.
   (For parts e and f, you must show your work to verify that you used the requested method)

13. According to *Bicycle Weekly News*, Davis, California is one of top ten bicycle friendly towns in America. In the town of Davis 64% of the population owns a bicycle. Suppose a random sample of thirty-five residents of Davis is selected and the town mayor is interested in calculating the probability of owning a bike.

    a. What is the random variable $X$?

    b. What is the distribution of $X$?

    c. Use JMP to find the CDF. Use the CDF to calculate the probability that between twenty and thirty residents sampled own a bicycle. (You must show your work to verify you used the requested method)

    d. How many of the residents sampled would you expect to see own a bicycle?

14. Suppose it is known that 86% of Americans make New Year's resolutions. Further suppose that Outliers Express has twenty-eight employees and the president of the company is interested in the number of his employees who made a New Year's resolution.

    a. What is the random variable $X$?

    b. What is the distribution for $X$?

    c. Use JMP to find the PMF. Use the PMF to calculate the probability that more than twenty-five employees of Outliers Express made a New Year's resolution. (You must show your work to verify that you used the requested method.)

    d. Find $\mu$.

    e. Find $\sigma$.

# Section 5

15. The fire marshal's office must inspect, on average, three buildings per day to meet regulatory requirements. Suppose the average number of buildings inspected per day is known to be 2.6. Suppose a building being inspected is independent per day with constant probability.

    a. What is the random variable $X$ represent?

    b. What is the distribution of the random variable $X$?

    c. Use JMP to find the PMF of $X$.

    d. Use JMP to find the CDF of $X$.

    e. Use the PMF to find the probability that the fire marshal's office meets regulatory requirements.

    f. Use the CDF to find the probability that the fire marshal's office exceeds regulatory requirements. (For parts e and f, you must show your work to verify that you used the requested method)

16. West Field County, Texas has been experiencing a mean of 2.6 rabid skunks per year. Find the probability that in a given year, there are more than four rabid skunks found. Suppose it is safe to assume that a skunk having rabies is independent with constant probability.

17. Suppose it is safe to assume that the number of accidents that happen at an assembly plant is independent with constant probability. Furthermore, suppose it is known that the average number of accidents per week is 1.4. Chad, the safety engineer, is interested in calculating probabilities about the number of accidents per week at the assembly plant.

    a. What is the random variable in this problem?

    b. What is the distribution of the random variable?

    c. Use JMP to find the PMF. Using the PMF calculate the probability of having at least 2 accidents per week. (You must show your work to verify that you used the requested method)

    d. Suppose Chad decides that he really is interested in calculating probabilities about the number of accidents per month (assuming 4 weeks equals one month). What is the distribution of this new random variable?

# Combining the Sections

18. Suppose a biologist is interested in monitoring the number of oak trees with oak wilt per acre. The average number of oak trees with oak wilt per acre is known to be 1.2. Further suppose the probability of a tree having oak wilt per acre is independent with constant probability.

    a. Find the probability that the biologist finds at most three trees with oak wilt on a one acre lot.

    b. How many oak trees with oak wilt would you expect to find on a 8-acre lot.

**19.** Suppose James has a box with eight balls in it; four red, four blue. Suppose James is going to randomly draw three balls without replacement from the box. Let the random variable $X$ is the number of red balls drawn. Find the PMF for $X$.

**20.** Of all the registered automobiles in Texas, 58% are sport utility vehicles. Suppose Carol's neighborhood has 20 vehicles registered.What is the probability that a majority of them are sport utility vehicles.

# CHAPTER 3 — CONTINUOUS PROBABILITY DISTRIBUTIONS

## ➡️ Section 1: The Probability Density Function, PDF:

The Probability Density Function, PDF, for a continuous random variable is the curve $f(x)$ that is found from smoothing out the histogram that is obtained through repeated sampling. For us, $f(x)$ will be given.

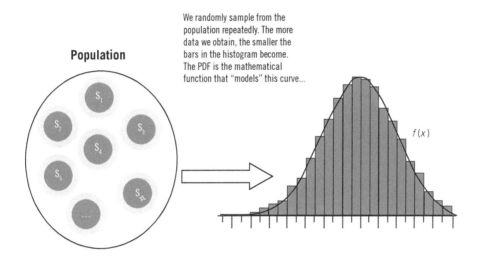

**Population**

We randomly sample from the population repeatedly. The more data we obtain, the smaller the bars in the histogram become. The PDF is the mathematical function that "models" this curve...

$f(x)$

- Properties:

    **1.** Probability is equal to area under the curve.

    **2.** $P(a < X < b)$ = area under the curve from $a$ to $b$ = _____ .

    **3.** The total area under the curve equals 1.

    **4.** $f(\text{x}) \geq 0$ for all $X$ in the range

    **5.** $P(X = a) = 0$

    Why?

    Answer:

- Example 1: Consider the following PDF, $f(x) = \frac{3}{4}x^2(2-x)$ where $0 < X < 2$.
  
  **a.** Verify that $f(x)$ is a PDF.

  **b.** Find $P(0.5 < X < 1)$

  **c.** Find $P(X \le 1.5)$

- Example 2: Suppose you have the following PDF for the continuous RV, $X$.

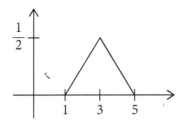

Find $P(X > 3)$

# Check Your Understanding

Suppose $f(x) = \dfrac{3}{50}(x^2 + 2x)$ with $1 < X < 3$ is the PDF for the random variable $X$.

**a.** Find $P(X > 0)$

**b.** Find $P(0 < X < 2)$

## ➡ Section 2: The Cumulative Distribution Function, CDF:

The Cumulative Distribution function, $F(x)$: The Cumulative Distribution Function gives cumulative probability. Since for a continuous distribution, probability is equal to area under the curve, the formula for finding $F(x)$ is:

$$F(x) = P(X \le x) = \int_{-\infty}^{x} f(y)dy$$

- Example:

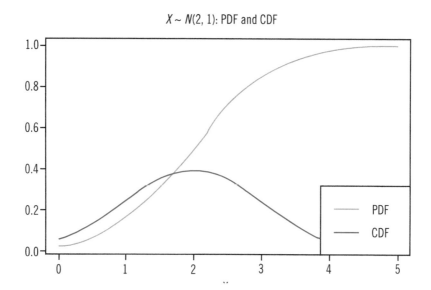

$X \sim N(2, 1)$: PDF and CDF

- Find $P(X \leq 2)$ using the PDF:

    Answer:

- Find $P(X \leq 2)$ using the CDF:

    Answer:

- Example 1: Consider the following PDF, $f(x) = \dfrac{3}{8}(49 - 14x + x^2)$, $5 \leq X \leq 7$.

    **a.** Find the CDF, $F(x)$.

**b.** Using the CDF, find $P(X \leq 6)$

**c.** Using the CDF, find $P(5.5 < X < 6.5)$

**d.** Using the CDF, find $P(X \geq 6.2)$

# Check Your Understanding

Suppose $f(x) = \dfrac{3}{50}(x^2 + 2x)$ with $1 < X < 3$ is the PDF for the random variable $X$.

**a.** Find the CDF, $F(x)$.

**b.** Use the CDF to find $P(X \leq 1)$

# ➡ Section 3: The Uniform Distribution:

Continuous random variables that have equally likely outcomes over their range of possible values have a Uniform Probability Distribution.

- The PDF is $f(x) = \frac{1}{d-c}$   $c \leq X \leq d$

- The graph of $f(x)$:

- The mean: $\mu = \frac{c+d}{2}$
- The standard deviation: $\sigma = \frac{d-c}{\sqrt{12}}$
- Recall, Probability = Area Under the Curve.

  Thus, $P(a < X < b)$ = area under the curve from $a$ to $b$

  =

Example: Suppose $X \sim Unif(2, 8)$. Find $P(X > 3)$:

# Check Your Understanding

Sue's Diner orders their chairs from Montana Table Company. When Sue calls Montana Table Company to place an order, she spends 5 to 30 minutes waiting on hold where the probability of how long someone waits on hold is constant for all minutes between 5 and 30. Suppose Sue is interested in calculating the probabilities about the length of time she waits on hold.

a. Find the PDF for $X$.

b. Find the probability that Sue spends between 10 and 20 minutes on hold.

# ➡ Section 4: The Normal Distribution: N(μ,σ):

- Characteristics of the Normal Distribution:

   **1.** The PDF is $f(x) = \dfrac{1}{\sqrt{2\pi}\,\sigma}\, e^{\frac{-(x-\mu)^2}{2\sigma^2}} \quad -\infty < X < +\infty$

   **2.** Symmetric bell-shaped curve

   **3.** Each curve is determined by the parameters μ and σ.
   μ is known as the location parameter; σ is known as the shape parameter.

   **4.** The Standard Normal Curve, $N(0, 1)$. Each normal curve can be standardized by using the formula:

   $$Z = \frac{x-\mu}{\sigma}$$

   The $z$-score, $Z$, is the standardized score of the random variable $X$ and describes for us how many standard deviations the random variable $X$ is above or below the mean.

   Example 1: Suppose the random variable $X$ = test scores $\sim N(80, 5)$

   What is the $z$-score for $X = 85$?

   Answer:

   What is the $z$-score for $X = 90$?

   Answer:

   What is the $z$-score for $X = 75$?

   Answer:

   What is the $z$-score for $X = 70$?

   Answer:

   What is the $z$-score for $X = 80$?

   Answer:

The *z*-score is just another way for representing the random variable $X$. In other words, I can tell you that you made a 95 on the test or that you made a standardized score of 3 and I am telling you the exact same thing.

The Normal Curve and the Empirical Rule:

- Example 1: Suppose $X \sim N(11, 2)$.

    **a.** $P(X < 12)$

    Answer:

    **b.** $P(6 < X < 13)$

    Answer:

    **c.** $P(X > 7.62)$

    Answer:

- A special notation with z-scores

  $Z_{\alpha}$: This notation represents the specific z-score that has area to its right equal to $\alpha$.

- Example 2: Find $Z_{0.05}$                      Find $Z_{0.95}$

- Example 3: Suppose again that $X \sim N(11, 2)$

  **a.** Find $x$ such that $P(X < x) = .025$

  Answer:

  **b.** Find $x$ such that $P(X > x) = .10$

  Answer:

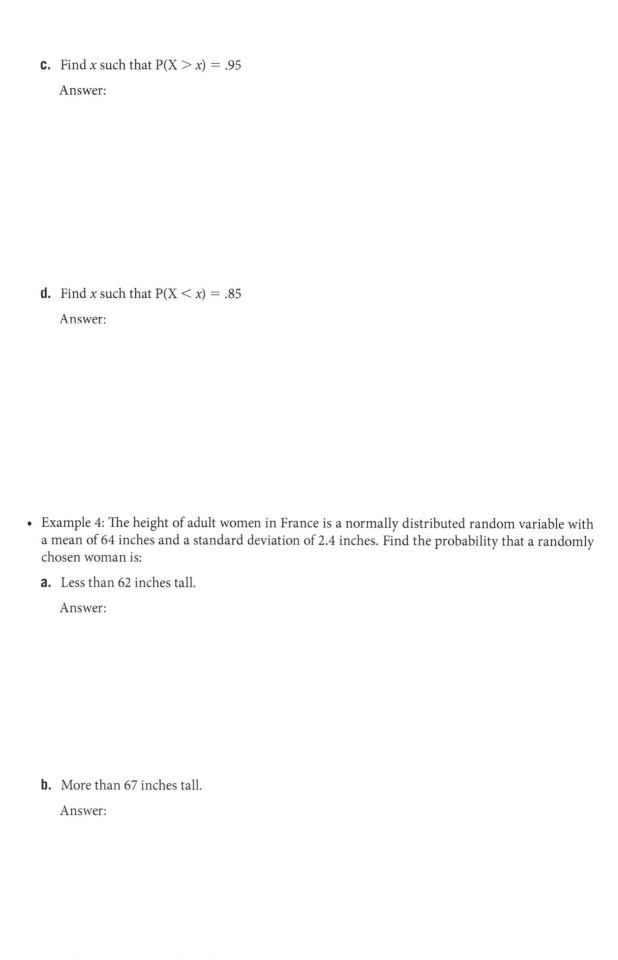

**c.** Find $x$ such that $P(X > x) = .95$

Answer:

**d.** Find $x$ such that $P(X < x) = .85$

Answer:

- Example 4: The height of adult women in France is a normally distributed random variable with a mean of 64 inches and a standard deviation of 2.4 inches. Find the probability that a randomly chosen woman is:

  **a.** Less than 62 inches tall.

  Answer:

  **b.** More than 67 inches tall.

  Answer:

**c.** Between 61 and 66 inches tall.

Answer:

**d.** If only 5% of the population of women in France is taller than Marie, what is her height in inches?

Answer:

- Example 5: The Triglyceride levels in healthy American adults are normally distributed with a mean of 134 mg/dL and a standard deviation of 14 mg/dL.

  **a.** If a healthy adult is randomly selected, what is the probability that their Triglyceride level is between 150 and 180?

  Answer:

  **b.** Physicians are told to recommend exercise and dietary changes if one's Triglyceride level is in the upper 3%. What is the cutoff for healthy adults?

  Answer:

# Check Your Understanding

1. Find $Z_{0.90}$

2. Find $Z_{0.30}$

3. The heights of cottonwood trees in Texas are normally distributed with a mean of 90 feet and standard deviation of 10 feet.

   a. Find the probability that a randomly selected pine tree is shorter than 75 feet.

   b. What is the probability that a randomly selected pine tree is between 70 and 105 feet tall?

   c. Find the probability that a randomly selected pine tree is taller than 97 feet.

   d. Find the height such that only 10% of trees are taller than that.

## ➡ Section 5: Continuous Probability Distribution Overview:

Probability Distributions— Describes _____ possible values that can be obtained by the random variable.

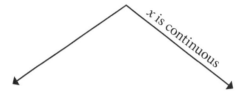

PDF: $f(x)$

probability = _____ under the PDF, $f(x)$, curve = _____ .

Total area = _____ .

$P(X = a) =$ _____ .

CDF: $F(x) =$

Probability is found by function evaluation.

$P(X \leq a) = P(X < a)$ _____ .

$P(X \geq a) = P(X > a) =$ _____ .

$P(a \leq X \leq b) = P(a < X < b) =$ _____ .

Uniform: all RV's have the _____ probability of occurring.

$f(x)$:

To find probability for a Uniform RV, find the _____ of the _____ .

Normal: $X \sim N(\mu, \sigma)$

Find probabilities by finding the z-score =

$P(Z < a) =$

$P(Z > a) =$

$P(a < Z < b) =$

# ⬤➤ Chapter 3 Homework:

## Section 1

**1.** Consider the following probability density function for the continuous random variable, $X$.

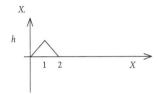

**a.** Find $h$

**b.** Find $P(X > 1)$

**c.** Find $P(X = \frac{1}{2})$

**2.** Let $X =$ temperature change in a controlled environment over a three hour period. The PDF, $f(x)$, for the random variable $X$ is $f(x) = cx^2$, $1 < X < 3$.

**a.** Find $c$

**b.** Find $P(X > 0)$

**3.** Consider the following probability distribution function for the continuous random variable, $X$.

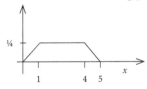

**a.** Find $P(2 < X < 4)$

**b.** Find $P(4 < X < 5)$

**4.** The elementary physical education teacher recorded the length of time $X$, in minutes, required for fourth graders to run one mile. The probability distribution is as follows:

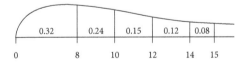

In parts (a) – (d), find the percentage of fourth graders that ran the mile in the specified time.

**a.** Less than 8 minutes

**b.** Between 8 and 12 minutes

**c.** More than 14 minutes

5. The length of time $T$, that a student spends in line at the university bookstore is described by the exponential pdf, $f(t) = 0.2e^{-0.2t}$ $t > 0$. What is the probability that a student will wait more than five minutes?

## Section 2

6. Given the following CDF, $F(x) = \begin{cases} 0 & X < -1 \\ \dfrac{x^3+1}{2} & -1 \le X \le 1 \\ 1 & X > 1 \end{cases}$

   a. Find $P(X \le \frac{1}{2})$

   b. $P(-2 < X < 0)$

   c. $P(-\frac{1}{2} < X < \frac{1}{2})$

7. Suppose $f(x) = 4x^3; 0 \le X \le 1$ and is zero for all other $X$.

   a. Find the CDF, $F(x)$

   b. Find $P(0 \le X \le 0.5)$

8. Let $X$ = tire tread thickness for all-season tires. The PDF, $f(x)$, for the random variable $X$ is

   $$f(x) = 600x^{-2}, \ 100 \le X \le 120.$$

   a. Find the CDF

   b. Find $P(108 \le X \le 112)$

   c. Find $P(X = 110)$

## Section 3

9. Suppose the probability of a continuous random variable $X$, occurring over the range from two to seven, is constant.

   a. Draw a picture of the PDF.

   b. Find the PDF function, $f(x)$.

   c. Find $P(3 < X < 6)$

10. Suppose Tommy is a delivery driver for Jed's Crispy Pizza Junction. It is known that the time it takes to deliver a pizza has a uniform distribution from twenty minutes to sixty minutes.

    a. What is the probability that the time to deliver a pizza is less than thirty-five minutes?

    b. If the time to deliver the pizza is longer than fifty minutes, the customer will receive $5.00 off the price of the pizza. What is the probability that the customer will receive $5.00 off the price?

# Section 4

11. Suppose $Z$ represents the standard normal random variable. If a value is selected at random from the standard normal distribution, find the probability that $Z$ is:

    a. Less than 0

    b. Between –2.35 and 1.65

    c. Greater than 1.86

    d. Greater than –2.15 and less than 1.67

    e. Less than –1.92 or greater than 2.41

12. Find the following $z$-scores:

    a. $Z_{0.05}$

    b. $Z_{0.95}$

    c. $Z_{0.15}$

    d. $Z_{0.80}$

13. Suppose $X \sim N(10,2)$.

    a. Find the probability that $X$ is at least 8.5.

    b. Find the probability that $X$ is at most 7.

    c. Find the probability that $X$ is between 6 and 14.

    d. Find $x$ such that $P(X < x) = 0.99$.

    e. Find $x$ such that $P(X > x) = 0.30$.

14. Suppose the incubation period for land turtle eggs is normally distributed with a mean of sixty-seven days and standard deviation 7.4 days.

    a. Find the probability that a land turtle egg will take at least eighty days to hatch.

    b. Find the probability that a land turtle egg will take at most fifty-five days to hatch.

    c. Find the probability that a land turtle egg will take between fifty and eighty-five days to hatch.

    d. Find the length of time such that 85% of the land turtle eggs will have hatched by then.

    e. Suppose a veterinarian becomes concerned about the health of the land turtle if it takes longer than ninety days to hatch. What is the probability of this happening if the turtle is healthy? Is it likely for this to happen?

    f. Find and interpret the 90$^{\text{th}}$ percentile.


15. Suppose the length of time of a whale's song is normally distributed with mean 12.5 minutes and standard deviation 2.6 minutes.

    a. Find the probability that a whale's song will last more than ten minutes.

    b. Find the probability that a whale's song will last between eight and sixteen minutes.

    c. Find the probability that a whale's song lasts less than five minutes or longer than twenty minutes.

    d. Find the length of time such that only 5% of whale's songs last longer than that.

    e. Find the length of time such that only 10% of whale's songs are shorter than that.


16. Let $X$ be the number of minutes after nine o'clock that the Downtown Marcela Street subway leaves the station. Assume that the distribution of times is approximately normal with mean fifteen minutes and standard deviation five minutes.

    a. If a person gets to the subway station at 9:10, what is the probability that the person has missed the Downtown Marcela Street subway?

    b. If a person is willing to risk a 15% chance of not making the Downtown Marcela Street subway, what is the maximum number of minutes after nine o'clock that the person can reach the station?

    c. What time should the person reach the station to have a 50% chance of catching the Downtown Marcela Street subway?


17. Suppose the random variable $X$ has a normal distribution with a mean of ninety and a standard deviation of six.

    a. Find the Quartiles

    b. Find $P_{30}$

# Combining the Sections

18. The recovery time taken for shoulder surgery for a professional pitcher is normally distributed with mean 7 months and standard deviation 1.5 months

    a. Find the probability that a randomly selected professional pitcher recovers in less than 6 months.

    b. What is the probability that a randomly selected professional pitcher recovers between 5 and 12 months?

    c. Find the length of time in months such that 10% of professional pitchers take less than that time to recover.

19. Given the CDF, $F(x) = \begin{cases} 0 & X < 1 \\ \dfrac{x^3 - 1}{26} & 1 \leq X \leq 3 \\ 1 & X > 3 \end{cases}$

    a. Find $P(X < 2)$

    b. Find $P(\frac{1}{4} < X < \frac{3}{4})$

20. Suppose the amount of weight in grams that a newborn gains in its first two weeks of life is uniformly distributed over the range of 30 grams to 420 grams. Find the probability that a newborn gains at least 100 grams in its first two weeks of life.

21. Suppose $f(x) = \dfrac{5}{282} (x^4 + 2x)$, $1 < X < 3$ is the PDF for the random variable, $X$.

    a. Find $P(X > 0)$

    b. Find the CDF.

**Probability Distribution** – describes _____ possible values
that can be obtained by the RV.

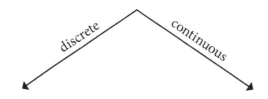

- **PMF:** $f(x) =$

| $x$ | $f(x)$ |
| --- | --- |
|  |  |

*Find prob. by asking,
which RV's satisfy the range*

**PDF:** $f(x)$: The mathematical function that is found by smoothing out the histogram that is obtained through repeated sampling.

*Probability =* _____ *under the PDF curve*

- **CDF:** $F(x) =$

| $x$ | $F(x)$ |
| --- | --- |
|  |  |

*Find prob. by drawing
Histogram and seeing what
mathematical manipulation
You should use*
i.e. $P(X \leq a) =$    or   $P(X < a) =$
or   $P(X \geq a) =$    or   $P(X > a) =$
or   $P(a < X < b) =$   or   $P(a \leq X \leq b) =$
or   $P(a \leq X < b) =$   or   $P(a < X \leq b) =$

**CDF:** $F(x) =$

*Probability is found by function evaluation.*

i.e.   $P(X \leq a) = P(X < a) =$
or   $P(X \geq a) = P(X > a) =$
or   $P(a < X < b) = P(a \leq X \leq b) =$

**Uniform:** all RV's have same probability
of occurring.

*Find probability by finding _____ of rectangle*

- **Binomial:** (5 questions). Use JMP to find the
PMF or CDF

- **Poisson:** (2 questions). Use JMP to find the
PMF or CDF

$N(\mu,\sigma)$**:** Find $z$-score.   Use $N(0,1)$ table to find
the area = probability

# SAMPLING DISTRIBUTIONS

CHAPTER 4

## ➡ Section 1: Sampling Distributions:

- Recall:

    - *Parameter*: A numerical value based on the _____ .

    - *Statistic*: A numerical value based on the _____ .

- We use the _____ to draw conclusions about the value

    of the _____ .

- Statistics are random variables, i.e., they will vary from sample to sample. Thus, they can be described by their probability distributions.

- *Sampling Distribution*: The sampling distribution is the distribution of all possible values that can be assumed by some statistic, computed from samples of the same size randomly drawn from the same population. In other words, it is the probability distribution of the sample statistic. The sampling distribution describes ALL possible values that can be assumed by the statistic.

- The sampling distribution describes the _____ .

    - Steps in constructing a sampling distribution:

        1. Randomly draw all possible samples of size *n* from a finite population of size *N*. (If population is infinite, then do "repeated sampling.")

        2. Compute the statistic of interest for each sample.

        3. List in one column the different observed values of the statistic, and in another column the corresponding frequency of occurrence of each distinct value of the statistic. *Or* Construct the histogram – the PDF will be the curve that is attained by smoothing out the histogram.

    - We will be interested in three characteristics about a given sampling distribution:

        1. mean

        2. variance

        3. shape (functional form)

# ➡ Section 2: The Sampling Distribution of x̄:

Many times we are interested in probabilities about a group instead of an individual random variable. Thus we need to know the distribution of the statistic. The statistic of interest right now is the sample mean.

- Recall:

    - _____ = population mean.

    - _____ = sample mean.

    - The difference between $X$ and $\overline{X}$:

        _____ = an _____ or a _____ object.

        _____ = the _____ of _____ randomly selected individuals or objects.

    - To draw conclusions about the population mean, we need to be able to calculate probabilities about the sample mean. Thus, we need to know the distribution of the sample mean. In other words, we need to construct the sampling distribution of $\bar{x}$.

- A Simulation for Constructing the Sampling Distribution of $\bar{x}$

    To illustrate the distribution of the sample mean, suppose we have the following population. Our random variable $X$ has a Beta(2, 2) distribution. The graph of the PDF is below. By looking at the graph, it is clear that although our population is symmetric, it is not normal.

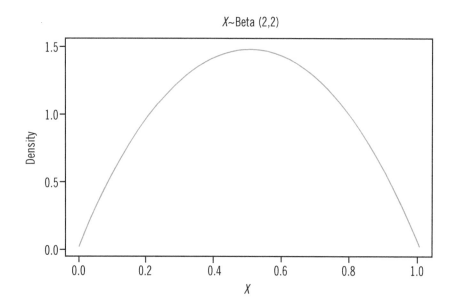

Following the steps for constructing sampling distributions, from this population, let's take 1000 samples of size 10 and for each sample, calculate $\bar{x}$. (This will give us 1000 $\bar{x}$ values.) This gives us the following graph:

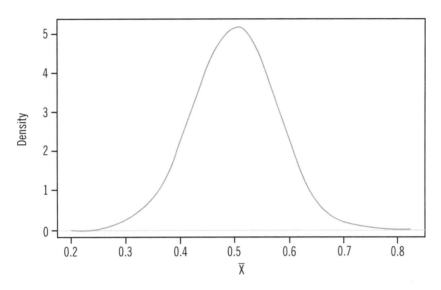

If we compare the graph of the population and the graph of the 1000 $\bar{x}$ values for $n = 10$, what do we notice?

Answer:

Now let's suppose we have the following population. Our random variable $X$ has an Exp($\frac{1}{2}$) distribution. The graph of the PDF is below. By looking at the graph, it is clear that our

population is most definitely not symmetric, in fact it is _____ and thus definitely not normal!

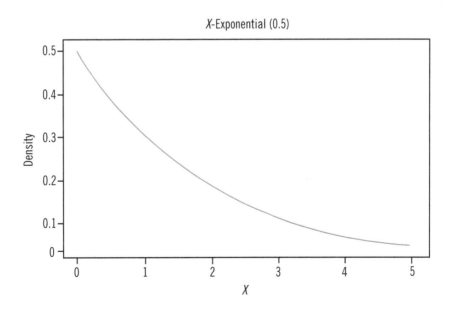

Again, following the steps for constructing sampling distributions, from this population, let's take 1000 samples of size 10 and for each sample, calculate $\bar{x}$. (This will give us 1000 $\bar{x}$ values.) This gives us the following graph:

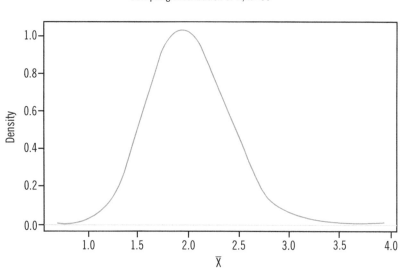

Sampling Distribution of $\overline{X}$, $n$=10

If we compare the graph of the population and the graph of the 1000 $\bar{x}$ values for $n = 10$, what do we notice?

Answer:


Now let's take 1000 samples of size 30 and for each sample, calculate $\bar{x}$. (This will give us 1000 $\bar{x}$ values.) This gives us the following graph:

Sampling Distribution of $\overline{X}$, $n$=30

If we compare the graph of the population and the graph of the 1000 $\bar{x}$ values for $n = 30$, what do we notice?

Answer:

Now let's take 1000 samples of size 150 and for each sample, calculate $\bar{x}$. (This will give us 1000 $\bar{x}$ values.) This gives us the following graph:

Sampling Distribution of $\overline{X}$, $n$=150

If we compare the graph of the population and the graph of the 1000 $\bar{x}$ values for $n = 150$, what do we notice?

Answer:

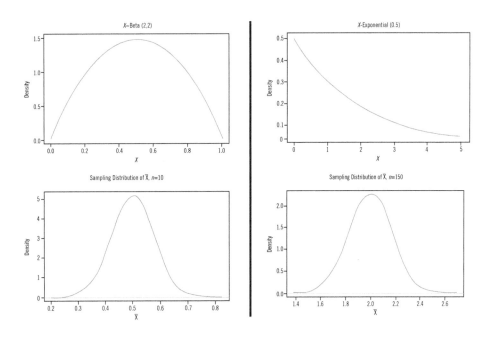

- Central Limit Theorem:

  In our sampling distribution simulations, we noted that as $n$ increases, the distribution of $\bar{x}$ becomes more bell-shaped. This happens whether the population distribution is symmetric or skewed. Thus, for a sufficiently large $n$, the distribution of $\bar{x}$ appears to be normally distributed. As it turns out, it *is* approximately normally distributed, which is given to us via the Central Limit Theorem.

- In Summation: (Properties of the sampling distribution of $\bar{x}$)

  **1.** The mean of the sampling distribution of $\bar{x}$: $\mu_{\bar{x}} = \mu$.

  **2.** The variance of the sampling distribution of $x$: $\sigma^2_{\bar{x}} = \sigma^2/n$

  **3.** What about the shape of the distribution of $\bar{x}$?

  *Theorem*: If a random sample of $n$ observations is selected from a normal population, the sampling distribution of $\bar{x}$ will also have a normal distribution.

  *Central Limit Theorem (CLT)*: Suppose a random sample of size $n$ is selected from *any* population. When $n$ is sufficiently large, the sampling distribution of $\bar{x}$ has an approximate normal distribution. As $n$ gets larger, this approximation becomes better.

- What does "sufficiently large" mean? Most elementary textbooks suggest the minimum is $n \geq 30$. *This is JUST a rule of thumb.* The required size of $n$ for $\bar{x}$ to be approximately normal depends on the population from which one took samples. If the population is highly skewed, then $n$ may need to be quite a bit more than thirty, possibly up to $n = 150$ (as we saw). If the population is symmetric and bell-shaped, then $n$ may be able to be much smaller than thirty, possibly as small as $n = 10$ (again as we saw). Thus, in practice, one should always look at a histogram of the data to verify normality, i.e., to verify that a sufficiently large enough sample was obtained.

- Examples

  **1.** On Texas Interstate Highways the speed of cars is a normally distributed random variable with mean of seventy mph and standard deviation of five mph.

    **a.** What does the random variable, $X$, represent?

**b.** Do we know the distribution of $X$?

**c.** What does the random variable, $\overline{X}$, represent?

**d.** Determine the sampling distribution of the sample mean for samples of size twenty.

**e.** What is the probability that the mean speed in a random sample of twenty cars is within two mph of the population mean speed of seventy mph.

**2.** Assume that cans of Dr. Pepper are filled so that the actual amounts have a mean of 12.00 oz. and a standard deviation of 1.2 oz.

**a.** What does the random variable, $X$, represent?

**b.** Do we know the distribution of $X$?

**c.** What does the random variable, $\overline{X}$, represent?

**d.** Determine the sampling distribution of the sample mean for samples of size thirty-six.

**e.** Find the probability that a sample of thirty-six cans will have a mean amount of at least 12.62 oz.

# Check Your Understanding

**1.** Suppose it is known that adolescents play an online game for an average of 81 minutes per day with standard deviation of 22 minutes. If a random sample of 40 adolescents is polled, find the probability that the mean minutes spent playing an online game per day is less than 92 minutes.

**2.** Suppose the life spans of exercise bicycles are known to be normally distributed with a mean of 8.25 years and a standard deviation of 3.25 years.

    **a.** What is the probability that the mean life span in a group of 10 randomly selected exercise bicycles is between 7 and 11 years?

    **b.** What is the probability that a randomly selected exercise bicycle has a life span of less than 13 years?

# ➡ Section 3: The Sampling Distribution of $\hat{p}$:

Sometimes we are interested in knowing the percentage of subjects that have a certain characteristic of interest. Or we might want to determine who the preferred presidential candidate will be in the upcoming election. In these cases our data is categorical, i.e., counting the number of subjects with the specified characteristic or counting the number of votes received by a candidate. For both examples, the parameter of interest is the population proportion.

- $p$ = population proportion.
- $\hat{p}$ = sample proportion.
- $\hat{p} = \dfrac{x}{n}$, where $x$ = the # of those with the characteristic and $n$ = the total sample size.
- The difference between $X$ and $p$ and $\hat{p}$

    _____ = the _____ (or objects) with the characteristic of interest.

    _____ = the _____ of the _____ with the characteristic of interest.

    _____ = the _____ of the _____ with the characteristic of interest.

- The Sampling Distribution of $\hat{p}$: To find the sampling distribution, we need to know the following:

    **1.** The mean of the sampling distribution of $\hat{p}$: $\mu_{\hat{p}} = p$

    **2.** The variance of the sampling distribution of $\hat{p}$: $\sigma_{\hat{p}}^2 = \dfrac{pq}{n}$

**3.** What about the shape of the distribution of $\hat{p}$?

Recall that when we are interested in the population proportion, our data generally consists of count data. In other words, we let our random variable $X$ count the number of subjects with the desired characteristic, etc. In order to determine the sampling distribution of $\hat{p}$, we define our random variable as such:

$$\text{Define } x_i = \begin{cases} 1 \text{ if the subject has the characteristic of interest} \\ 0 \text{ if the subject does not have the characteristic of interest.} \end{cases}$$

Thus we can define $\hat{p}$ to be

$$\hat{p} = \frac{\sum x_i}{n}$$

- By looking at the formula for $\hat{p}$, what can we say about the functional form of $\hat{p}$?

  Answer:

- What are the assumptions?

  Answer:

- Examples

  **1.** Mayor Alice Cook is running for reelection and believes that she can win the election if she can poll a majority of the votes in the Northern precinct. She also believes 55% of the city's voters favor her. If 150 voters show up in the Northern precinct, what is the probability Mayor Alice Cook receives a majority of the Northern precinct votes?

**2.** The Texas High School Coaches Association claims that 12% of high school athletes have suffered a career ending injury while playing a high school sport? Out of 100 randomly chosen high school athletes, what is the probability that less than 7% will have a career ending injury while playing a high school sport?

The Relationship between $\hat{p}$ and the Binomial Distribution:

**3.** The American Restaurant Association finds that 8% of persons who make reservations do not show up. Suppose Andy's Five-Star Cuisine Restaurant has 160 tables and reservations are made for all tables for Friday night. Further suppose that ten couples, who don't have reservations, show up to eat. What is the probability that all the couples will be able to eat at their own table?

- Example 3 illustrates for us that a binomial random variable for large *n*, is approximately normal. Below is the histogram of the random variable from example 3.

Notice that the histogram for the binomial distribution above fits quite nicely under the normal curve. Thus, when our RV has a Binomial distribution, if the sample size is sufficiently large, we can convert the *x* to a $\hat{p}$ and approximate the problem using the normal distribution.

## Check Your Understanding

1. Suppose that 35% of all high school students take shop class. If you sample 10 high school students, what is the probability that more than 30% of the students you sample take shop class?

2. Suppose that 30% of all retired people take an online class. If you sample 100 retired people, what is the probability that more than 35% of the them are taking an online class?

## ➡ Section 4: Probability Distributions Overview:

When you see the key word "probability," your first question should be, "Is my probability about a random

variable *X* or about a statistic?" If it is about *X*, then we are interested in a _____ or

_____. If it is about a statistic, then we are interested in a _____

and thus given a _____.

- *X?*

- *A Statistic?*

- *X?*

# ➡ Chapter 4 Homework:

## Section 2

1. Suppose the random variable $X$ is drawn from a population with mean thirty and standard deviation five.

   a. For samples of size sixty, what is the distribution of the sample mean?

   b. Is the answer to part *a* dependent upon the sample size?

2. Suppose the random variable $X$ is drawn from a normal population with mean fifty and standard deviation two.

   a. For samples of size sixty, what is the sampling distribution of the sample mean?

   b. Is the answer to part *a* dependent upon the sample size?

   c. Is the answer to part *a* dependent upon knowing the distribution of $X$?

3. The average arm span for females is 63.75 inches with standard deviation 3.4 inches. Certain athletes, such as swimmers, rowers, or boxers benefit from having a large arm span. Suppose forty female swimmers are randomly selected. What is the probability that their average arm span is greater than 65.8 inches?

4. According to the American Music Industry, the length of a song is normally distributed with a mean of 240 seconds and standard deviation thirty seconds. Suppose a random sample of fifteen songs is selected from the morning program. What is the probability that the mean length of the song is between 225 and 260 seconds long?

5. The recovery time taken for shoulder surgery for a college fast-pitch pitcher is normally distributed with mean 8 months and a standard deviation of 1.2 months. If you randomly select 16 college fast-pitch pitchers, what is the probability that their mean recovery time is less than 8.5 months?

# Section 3

**6.** A manufacturer of cell phones purchases batteries from a vendor. It is not uncommon to receive defective batteries from this vendor, thus whenever a shipment is received, a sample of batteries is inspected. For this month's shipment, a random sample of 280 batteries is selected and each battery is inspected. If the proportion of defective batteries sampled is more than 1.5%, the entire shipment will be returned to the vendor. What is the probability that this month's shipment will be returned if the true proportion of defective batteries is 4%?

**7.** According to the Federal Insurance Agency statistics, the proportion of teenage drivers between the ages of 16 to 19 who have received a speeding ticket is 54%. From a random sample of one hundred teenage drivers, what is the probability that a majority of them have received a speeding ticket?

**8.** The proportion of college students that participate in a work-study program is 60%. From a random sample of fifty college students, what is the probability that at least forty of them participate in a work-study program?

# Combining the Sections

**9.** Fifty-one percent of adults in the United States whose New Year's resolution was to exercise more achieved their resolution. Suppose a random sample of sixty-five American adults whose New Year's resolution was to exercise more is chosen. What is the probability that a majority of them achieved their resolution?

**10.** Fifteen percent of adults in the United States do not make New Year's resolutions. Suppose a random sample of sixteen American adults is chosen. What is the probability that a majority of them do not make New Year's resolutions?

11. The time in seconds that it takes a Baylor student to jog from the Student Union Building to the Student Life Center is distributed as a Normal random variable with mean ninety-five seconds and standard deviation of fifteen seconds.

   a. What is the probability that a student runs to the Student Life Center in at most two minutes?

   b. What is the time at which the fastest 20% of students will have completed the run?

   c. Suppose fifteen students decide to race from the Student Union Building to the Student Life Center. What is the probability that the average time it takes the students to get to the Student Life Center exceeds 100 seconds?

   d. Suppose fifteen students decide to race from the Student Union Building to the Student Life Center. What is the probability that the average time it takes the students to get to the Student Life Center exceeds 150 seconds?

12. Suppose that 21% of men in the United States consider fishing their favorite leisure-time activity. A random sample of ten men is selected. What is the probability that at least 20% of them will say that fishing is their favorite leisure-time activity?

13. The incubation period for the eggs of a house wren is normally distributed with a mean of 336 hours and a standard deviation of 3.5 hours.

   a. Find the probability that an egg has an incubation period between 328 and 338 hours.

   b. Suppose a researcher has fifteen house wren eggs in an incubator. Find the probability that the average incubation time for the fifteen eggs is greater than 337 hours.

   c. Suppose it is known that 33% of House Wren eggs never hatch. Suppose an ornithologist has forty house wren eggs. Find the probability that less than 25% never hatch.

# CHAPTER 5
# ESTIMATION AND CONFIDENCE INTERVALS FOR ONE PARAMETER

*Statistical Inference* – the process of making judgments about a population based on properties of a sample randomly drawn from the population.

*Estimation* – involves approximating the value of an unknown parameter.

*Hypothesis Testing* – involves choosing between two opposing statements concerning a parameter

## ➡ Section 1: Point Estimate:

Point estimate – when we use a statistic to estimate the value of a parameter.

- Suppose I ask one hundred randomly selected Baylor students how much money they spent eating out last month. The sample mean for the one hundred students is $182.25.

  If I make the following statement: "*A random sample of 100 Baylor students spent $182.25 eating out last month*," am I describing the population or the sample?

  Then I have simply made a _____ statement.

  If, however, I make the statement; "*Baylor students spent an average of $182.25 eating out last month,*"am I now describing the population or am I still describing the sample?

  Then I have made an _____ statement. I am using information from the

  _____ to describe the _____.

  In other words, I am using the _____ as an estimate of the

  _____.

- We define the sample mean, $\bar{x}$, to be a *point estimate* of the population mean, $\mu$.
- We define the sample proportion, $\hat{p}$, to be a *point estimate* of the population proportion, $p$.

In the above example, I am interested in the *typical* amount Baylor students spend on eating out per month. Recall, this is what we refer to as a Measure of Central Tendency. What are the three Measures of Central Tendency?

Answer:

Recall that sometimes it is best to use one statistic versus another when describing the typical data value. How do we know which statistic to use? We know that if our data is skewed or contains outliers, we should use the _____. However, the mean is the most commonly one used because of the following:

- When choosing which statistic to use to estimate the parameter, we want to choose the statistic that has the following two properties:

  **1.** The statistic is *unbiased.*

  **2.** The statistic has *minimum variance.*

     Let's first define what we mean by unbiased: Consider the following distributions:

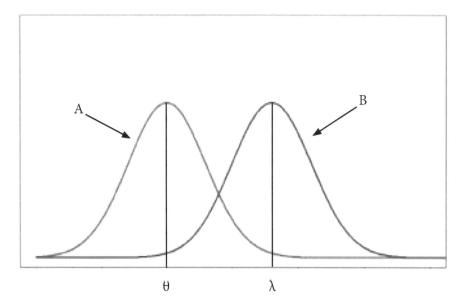

- Assume $\theta$ is the population parameter that we wish to estimate. Also, assume both $\hat{\theta}$ and $\hat{\lambda}$ are estimates of $\theta$.

- Let A be the distribution of $\hat{\theta}$. Thus, the curve A describes:

  Answer:

- Let B be the distribution of $\hat{\lambda}$. Thus, the curve B describes:

  Answer:

- By looking at the curve A, we know the $E[\hat{\theta}] =$ _____. This implies:

  Answer:

- By looking at the curve B, we know the $E[\hat{\lambda}] =$ _____. This implies:

  Answer:

- The statistic $\hat{\theta}$ is *unbiased* if the mean of the sampling distribution of $\hat{\theta}$ is equal to the parameter, $\theta$. i.e., $\hat{\theta}$ is an unbiased estimator of $\theta$ if $E[\hat{\theta}] = \theta$.

- We say $\hat{\lambda}$ is a *biased* estimator of $\theta$, but an unbiased estimator of _____.

- Let's now define what we mean by minimum variance. Consider the following distributions:

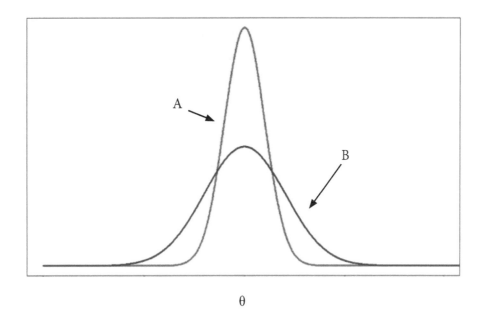

$\theta$

- Let A be the distribution of $\hat{\theta}$ and B be the distribution of $\hat{\delta}$.

- $E[\hat{\theta}]$ and $E[\hat{\delta}] =$ _____, thus both $\hat{\theta}$ and $\hat{\delta}$ are _____ estimators of $\theta$.

- Label on the graph which range of values of $\hat{\theta}$ are most likely to occur.

- Label on the graph which range of values of $\hat{\delta}$ are most likely to occur.

- The statistic $\hat{\theta}$ said to have *minimum variance* if the sampling distribution of the statistic has the smallest variance of all the unbiased estimators.

Why is this good thing?

Answer:



Now back to our Baylor eating out example:

Do you think it is very likely that the true value of the population mean is $182.25? In other words, do you think it is true that the average amount that ALL Baylor students spent on eating out last month is $182.25? Most likely not. It *could* be the case that the one hundred students selected just happened to spend more on eating out than the other Baylor students. In fact, the average for all Baylor students (the *population mean*) could be very different from the *sample mean* of $182.25.

Unless we perform a census, we can never know with absolute certainty even approximately what the true population mean is. For instance, what if one student not polled happened to spend $1,000 on eating out last month? We know this outlier will drastically affect the sample mean!

Can we improve our estimate of the population mean? **Yes!**

Instead of just using a point estimate, can we do better? **Yes!**

# ▶ Section 2: Confidence Intervals:

A *confidence interval* is an interval estimate of the parameter in which we include how "confident" we are that the interval contains the parameter we are estimating.

Instead of using one point to estimate the parameter, we use a range of values as the estimate, making sure we include how "confident" we are that the interval contains the true value of the parameter.

- What do we mean when we talk about how "confident" we are that the range contains the parameter?

A 95% confidence interval is an interval generated by a process that's right 95% of the time. Similarly, a 90% confidence interval is an interval generated by a process that's right 90% of the time and a 99% confidence interval is an interval generated by a process that's right 99% of the time.

Another way to say this is that if we were to replicate our study many times, each time reporting a 95% confidence interval, then 95% of the intervals would contain the true value of the parameter.

As an illustration, suppose we are drawing from a population in which we know the population mean is equal to fifty. We are going to construct ten different 90% confidence intervals from this population. The picture below illustrates what we mean by 90% confidence.

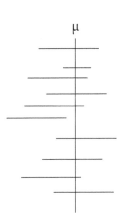
μ

Two things to notice:

**1.** How many of the confidence intervals contain the true value of μ? How many do not?

**2.** Are the endpoints for each confidence interval the same or different?

- Comments:

  **1.** If we construct a 90% confidence interval for μ to be, *5 < μ < 15*, does that imply that 90% of the time, the value of μ is between *5* and *15*?

  Answer:

  **2.** In practice, we perform our study only once. i.e., we construct one confidence interval. We have no way of knowing whether our particular interval is correct, but we behave as though it is.

  **3.** In theory, we can construct intervals of any level of confidence from 0 to 100%. However, the amount of confidence affects the length of the interval. *Keeping all else the same, if we increase the confidence of the interval, we increase the length of the interval.* The longer the interval, the more likely it is to contain the quantity being estimated, but the length could be too long to be useful in estimation.

  **4.** Ninety-five percent has been found to be a convenient level for conducting scientific research, so it is used almost universally. Intervals of lesser confidence would lead to too many misstatements. Greater confidence would require more data to generate intervals of usable lengths.

- How do we construct a confidence interval (CI)?

  When constructing a CI, we begin with a point estimate and "stretch" that point estimate into an interval estimate (or range of values) in a way that allows us to quantify our "uncertainty" about that interval containing the true value of the parameter.

  The amount in which we stretch the point estimate is called the Margin of Error and denoted by *E*.

  This gives us the following general formula for finding a CI.

  $$\text{Pt. Est.} \pm E$$

  - When estimating a parameter, what "type" of point estimate do we want to use?

    Answer:

- What statistical tool do we use to quantify our uncertainty?

  Answer:

- What describes for us all possible values of a point estimate/statistic?

  Answer:

Thus, when calculating a CI, we are finding the points $a$ and $b$ on the distribution curve of the point estimate such that the probability of being between $a$ and $b$, centered about the mean of the distribution, is equal to our level of confidence.

Two points to understand:

1. The curve we are using is the _____ curve of the point estimate and describes for us _____ possible values that can be obtained by the

   _____ .

2. The mean of the distribution is equal to the _____ we are estimating.

- The picture below illustrates this.

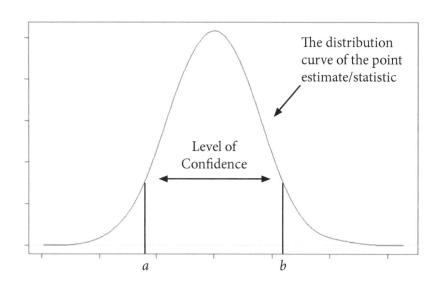

# ➡ Section 3: A (1 − α)% Confidence Interval for *p*:

Review:

- What is the point estimate of *p*?

- What is the sampling distribution (including assumptions)?



- Thus, we know the _____ curve centered about _____

  describes for us _____ possible values of _____ .

- If we want to find the points $a$ and $b$ on this curve, what must we find first?




- If I want to find a 90% CI, the _____ for the points $a$ and $b$ would be?




- If I know the points $a$ and $b$ are _____ standard deviations above and below the mean of the distribution then what do I do to determine how many units the points $a$ and $b$ are above and below the mean of the distribution?




- The formula for finding a 90% CI for $p$ is:




- What is wrong with this formula? What is the goal for finding a CI?




- Two ways to handle this problem:

- Hence, what is the general form of the CI for $p$?

- Example 1: A poll was taken of 1200 U.S. nurses. The nurses sampled were asked whether they have ever administered an incorrect amount of medicine to a patient and 240 responded "yes." Use these data to obtain a 95% confidence interval for the proportion of all US nurses who have made an error in administering medicine to a patient.

- Interpretation:

- Margin of Error, $E$:

- Interpretation of $E$:

- Can we be 95% confident that more than 20% of US nurses have made an error in administering medicine?

- Can we be 95% confident that less than 25% of US nurses have made an error in administering medicine?

- For a CI to support a claim, _____ value within the CI has to support the claim!

- Let's make sure we really understand what a CI implies and/or doesn't imply. Answer the following True/False Questions:

  1. The above implies that 95% of the time the proportion of US nurses who have made an error in administering medicine is between 17.74% and 22.26%.

     A  True                          B  False

  2. The above implies that $P(.1774 < p < .2226) = 0.95$

     A  True                          B  False

  3. Saying we are 95% confident implies that the method used to construct the interval will result in an interval, 95% of the time, which contains the true proportion of US nurses who have made an error in administering medicine.

     A  True                          B  False

  4. We have no idea if the true proportion of US nurses who have made an error in administering medicine is between 17.74% and 22.26%, but we are 95% confident that it does, therefore we use these as valid estimates of the true value of $p$.

     A  True                          B  False

# Check Your Understanding

Suppose the administrator in charge of student tickets is interested in estimating the proportion of students who left before half time at the last football game. This administrator randomly selects 60 students who attended the last game and asks if they left before half time. Of the 60 students, 32 said yes, they did.

    **a.** Find a 95% confidence interval for the true proportion of students who left before half time at the last football game.

    **b.** Can the administrator claim with 95% confidence that a majority of the students left before half time of the last game? Explain your answer.

# ➡ Section 4: Determining *n*:

In theory a researcher can construct a CI at any level of confidence. But, we also know that holding all else the same, if we increase the level of confidence, the length of the CI will increase. Our goal is to have the maximum level of confidence with the minimal (usable) length.

In many studies, the researcher wants to be able to estimate the parameter with a specific level of confidence and to be within a specified value. In order to do this, the researcher needs to be able to determine the sample size needed to insure this certain degree of accuracy in estimation.

There are several factors to consider in doing this:

    **1.** *Level of Confidence* – refers to how "confident" we are that the interval actually contains the true value of the parameter.

    **2.** *Precision* – refers to the width of the confidence interval.

    **3.** *Variability* – refers to the variability of the population. The more variability in the population the larger the sample size needs to be.

    **4.** *Complexity of Analysis* – the more variables that one is analyzing simultaneously, the larger the sample size needs to be.

    **5.** *Resources* – larger sample sizes require more time and money.

- *Determining the sample size needed when estimating a proportion:*
  - What is the formula for constructing a CI for *p*?

    Answer:


  - Which part of the Margin of Error refers to the precision and the level of confidence?

    Answer:

- How does the precision relate to the Margin of Error?

    Precision = _____.

    Margin of Error = _____.

- What about $\hat{p}$ and $\hat{q}$? Do we know their value? What do we do if not?
  Answer:

- Formula for finding $n$ when estimating $p$:
  Answer:

- Example 1: The nursing manager was concerned to see the proportions found in the confidence interval above. She decides to calculate her own confidence interval. What sample size should she take so that she can estimate the true proportion of nurses who have made an error in administering medicine to be within 2% with 99% confidence?

    **a.** Suppose the point estimate of 20% from the previous example is valid.

    **b.** Assume there is no information known about the true proportion.

# Check Your Understanding

Suppose it is believed that 36% of homes in the United States subscribe to some type of home viewing movie service. We would like to estimate the true proportion of homes in the US that subscribe to some type of home viewing movie service. What sample size should we use to ensure we estimate the truth within 3% with 95% confidence?

## ➡ Section 5: A (1 – α)% Confidence Interval for μ:

- The general form for a confidence interval is as follows:

    Point Estimate ± Margin of Error

    Point Estimate ± Reliability Factor * Standard Error

- So for μ, we have the following general form of the CI:

    Answer:

  - What is the goal when finding a CI for μ?

    Answer:

  - Do we know the value of μ?

    Answer:

  - How likely is it that we will know the value of σ?

    Answer:

- When we replace σ with *s*, the distribution of the point estimate is no longer normal. It was discovered by Dr. Gosset to have a student's *t*-distribution (*t*-distribution)  Thus, we have the following formula for finding a CI for μ:

    Answer:

- Properties of the student's *t*-distribution:
  1. symmetric bell-shaped curve
  2. centered at the mean = 0
  3. has more variability than the $N(0,1)$ curve. (fatter tails)
  4. there are infinitely many different *t*-curves. Each one is determined by its degrees of freedom.
  5. as $n \to \infty$, the *t*-curve $\to N(0,1)$.

- In order to use the student's *t*-distribution, our population must be normally distributed. We call this a distributional assumption. Thus, the assumption is that the population from which the data is drawn from must be normally distributed.

- How do we check normality? We have a computer statistical package, such as JMP, create a Normal Quantile Plot, (also called a QQ-Plot).

This is what the QQ plot looks like for data that is normally distributed:

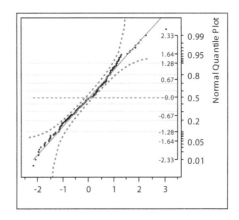

If the data values are forming an overall concave shape, either concave up or concave down, then we cannot assume that the population is normally distributed. In fact, the distribution is skewed.

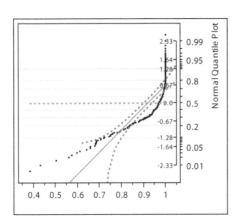

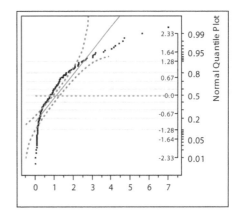

Over-all concave up implies the population is _____.

Over-all concave down implies the population is _____.

- Example 1: Crime analysts compile data on misdemeanor theft crimes. A theft is considered a misdemeanor if the item stolen is not considered too valuable. A sample of twenty-four of last year's misdemeanor thefts had a mean value loss of $53.63 with standard deviation $25.53.

    **a.** Obtain a 95% confidence interval for the mean value lost of last year's misdemeanor theft crimes.

    **b.** Interpret the CI

    **c.** Find the margin of error, $E$:

    **d.** Interpretation of $E$:

- Suppose the data is given. If we have the data, we can use JMP to find the CI and to check our assumption.

| | | | | | | | |
|---|---|---|---|---|---|---|---|
| 47 | 34 | 28 | 80 | 96 | 43 | 76 | 55 |
| 27 | 14 | 106 | 75 | 57 | 29 | 27 | 45 |
| 88 | 39 | 89 | 54 | 48 | 43 | 65 | 22 |

- JMP Instructions:

5. Input data

6. Under analyze, choose distribution

7. Click on the column containing the data, then click on *Y*, then hit okay.

8. Under the red options button next to distribution, choose stack

9. Under red options button next to column 1 (or whatever you named your column), choose Normal Quantile Plot

10. Under red options button next to column 1 (or whatever you named your column), choose confidence interval

11. Choose the correct significance, hit okay.

- JMP Output:

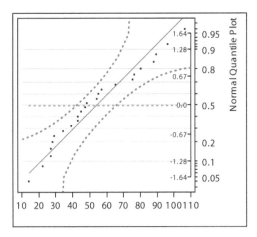

**Confidence Intervals**

| Parameter | Estimate | Lower CI | Upper CI | 1-Alpha |
|---|---|---|---|---|
| Mean | 53.625 | 42.84367 | 64.40633 | 0.950 |
| Std Dev | 25.53227 | 19.84403 | 35.81566 | 0.950 |

# Check Your Understanding

1. Suppose it was known that people talked on their home phone for an average of 21 minutes per day. Researchers want to know if having such easy access to cell phones has changed the amount of time people talk on a phone. Thus, they would like to estimate with 95% confidence the average number of minutes that people talk on a phone (cell or land). Suppose a random sample of 40 people is polled, and the mean number of minutes talked on a land line or cell phone is 24 minutes with standard deviation 5 minutes.

   a. Estimate with 95% confidence the average number of minutes that people talk on a phone (cell or land)

   b. Can you be 95% confident that the average number of minutes people talk on a phone (cell or land) is more than 21 minutes? Explain your answer.

2. Suppose a city architect is interested in the heights of the buildings in his town. He would like to know if the average height of the buildings in his town is different than 105 ft. The architect randomly selects 10 buildings and their heights, in feet, are listed below.

   | 106 | 109 | 107 | 108 | 112 | 113 | 104 | 105 | 115 | 114 |
   |-----|-----|-----|-----|-----|-----|-----|-----|-----|-----|

   a. Find a 99% confidence interval for the average height of the buildings.

   b. Can the city architect conclude with 99% confidence that the average height is different than 105 ft.

## ➡ Section 6: Estimation and Confidence Interval Overview:

Inferential Statistics ⟶    Drawing conclusions about the _____ based

on the _____ .

i.e., Using the _____ as an estimate for the

_____ .


Estimation ⟶    estimating the value of a _____ .

Point Estimate: Using one point (one value) to estimate the _____ .

_____ is a point estimate for $\mu$.

_____ is a point estimate for $p$.

Confidence Interval ⟶ an interval estimate for the _____ where

we include how confident we are that the interval contains

the _____ we are attempting to estimate.

General form of a CI: pt. est. $\pm E$ ⟶ $(\underline{pt.est. - E}, \underline{pt.\ est. + E})$

CI on $p$:

CI on $\mu$:

# ➡️ Chapter 5 Homework:

## Section 1

1. When choosing a point estimate, we want to choose the point estimate that has what two properties?

2. Is $\bar{x}$ an unbiased estimator of $\mu$? Why or why not.

3. Is $\hat{p}$ an unbiased estimator of $p$? Why or why not.

4. What is the point estimate for $p$?

5. What is the point estimate for $\mu$?

6. The ASA softball association is interested in knowing the average pitching speed of a 12U fastpitch softball pitcher. A random sample of forty pitches is recorded. The average speed of the pitch is found to be forty-one mph with standard deviation two mph.

    a. What is the parameter of interest?
    b. What is the point estimate you would calculate to estimate the parameter? What is its value?

7. Across the nation, many college students have been complaining to the dean of students about the temperature in the classrooms being at an "uncomfortable level", either too cold or too hot. Baylor's dean of students is interested in determining the true proportion of Baylor students that feel the classroom temperature is at an uncomfortable level. Out of a random sample of 200 Baylor students, ninety-eight say that the classrooms are either too hot or too cold.

    a. What is the parameter of interest?
    b. What is the point estimate you would calculate to estimate the parameter? What is its value?

# Section 2

**8.** What is the goal of a confidence interval?

**9.** Suppose your dorm refrigerator is no longer working and you want to buy a new one. Your roommate thinks it would be cheaper to have yours repaired. To settle the argument, a random sample of ten dorm refrigerators is taken and the resulting 95% confidence interval for the mean repair cost is (59.48, 90.52). Based on the confidence interval, can you be 95% confident that it will cost more than $62 to repair the dorm refrigerator?

**10.** Suppose the margin of error for estimating a parameter is ten. What is the length of the resulting confidence interval?

**11.** Suppose the length of a confidence interval is eight. What is the Margin of Error?

**12.** Suppose the length of a confidence interval is ten. If the point estimate is forty-five, what is the confidence interval?

**13.** Suppose the margin of error of a confidence interval is six. If the point estimate is eighty, what is the confidence interval?

**14.** Keeping all else the same, is a 95% confidence interval narrower or wider than a 90% confidence interval?

**15.** Keeping all else the same, is a 95% confidence interval for $n = 100$ narrower or wider than a 95% confidence interval for $n = 200$?

**16.** A 95% confidence interval for the proportion of Americans under the age of thirty who skip breakfast is found to be (0.2025, 0.2795). For each choice below, answer True or False:

    **a.** T  or  F   There is a 95% probability that the true proportion of Americans under the age of thirty who skip breakfast is between 20.25% and 27.95%.

    **b.** T  or  F   If this study were to be repeated many times with samples of the same size, then 95% of the time the true proportion will be between 0.2025 and 0.2795, on average.

    **c.** T  or  F   We can be 95% confident that, on average, the proportion of Americans under the age of thirty sampled who skip breakfast is between 0.2025 and 0.2795.

    **d.** T  or  F   We can be 95% confident that, on average, the proportion of Americans under the age of thirty who skip breakfast is between 0.2025 and 0.2795.

    **e.** T  or  F   If we were to construct one hundred confidence intervals, then 95% of them would contain the true proportion of Americans under the age of thirty who skip breakfast, five percent would not.

# Section 3

**17.** One of the hot topics in the last presidential election was with respect to our citizen's health and whether or not the government has a right to limit our choices in what we eat or drink. A random sample of 1046 Americans was obtained. The question asked was "Do you support our government banning specific types or sizes of food or drinks?" Of those surveyed, 895 responded, "no."

    **a.** Estimate with 95% confidence the proportion of Americans who do not support the government banning specific types or sizes of food or drinks.

    **b.** Using the confidence interval found in part a, can you conclude that a majority of Americans do not support the government banning specific types or sizes of food or drinks. Why or why not.

**18.** The governor of New York has banned selling or purchasing "large" soft drinks, where a "large" is defined as being more than 16 oz. Sonic, famous for its Route 44 drinks is interested in estimating the proportion of New Yorkers that favor this ban. From a random survey of 800 New Yorkers, 375 answered no to the question, "Do you favor the governor's ban on 'large' soft drinks?"

    **a.** Find a 97.5% confidence interval for the proportion of New Yorkers who favor the ban on "large" soft drinks.

    **b.** Interpret the confidence interval.

    **c.** Can you conclude with 97.5% confidence that a majority of New Yorkers favor the ban on "large" soft drinks? Why or why not.

    **d.** Without doing any other calculations, can you be 99% confident that a majority of New Yorkers favor the ban on "large" soft drinks?

# Section 4

**19.** What sample size should a researcher take if he wants to estimate a population proportion to be within 2% of the true value with 90% confidence?

**20.** What sample size should a researcher take if he wants to estimate a population proportion to be within 3% of the true value with 95% confidence? Previous research suggests the proportion is equal to 25%.

**21.** To estimate the proportion of teen traffic deaths in Texas last year that was caused by speeding, determine the necessary sample size for the estimate to be accurate to within .05 with probability .99.

    **a.** Suppose from a previous study, we expect the proportion to be about .30.

    **b.** Suppose we do not have any prior knowledge about the true proportion.

# Section 5

**22.** With economical concerns being so great, many people are no longer purchasing new vehicles, but are instead keeping their vehicles for a longer period of time, until they no longer work. The American Automobile Association is interested in estimating the average age of vehicles for Americans. A random sample of thirty-five Americans was obtained. The average age of their vehicles was 6.2 years with standard deviation 1.2. Estimate with 99% confidence the average age of vehicles for Americans.

**23.** Many people have heard of the "Freshmen 20," referring to the belief that many college freshmen will gain 20 lbs. their freshmen year in school. Administrators were interested in knowing if this is really true. A random sample of thirty freshmen (at the end of their freshmen year) was obtained. The average amount of weight gained by the freshmen is listed below.

|      |      |      |      |      |      |      |      |      |      |
|------|------|------|------|------|------|------|------|------|------|
| 21.8 | 20.6 | 21   | 20.2 | 19.2 | 17.8 | 22.8 | 20.2 | 22   | 19.4 |
| 18.6 | 18.2 | 20   | 19   | 21   | 23   | 19.6 | 19   | 24.6 | 19.6 |
| 17.8 | 20.6 | 20.8 | 17.2 | 22.4 | 19.2 | 23.4 | 18.2 | 20.6 | 21.4 |

    **a.** Find a 97.5% confidence interval for the average amount of weight gained for freshmen.

    **b.** Interpret the CI.

    **c.** Using your CI found in part a, can you conclude with 97.5% confidence that college freshmen gained more than 20 lbs.?

# Combining the Sections

**24.** Does exercise lower a person's resting pulse rate? This is a question researchers are interested in investigating. Let $X$ be the pulse rate of fifteen randomly selected individuals who are regular attending members of a local gym. The mean of those fifteen pulse rates is 69.95 with standard deviation 7.67.

    **a.** State the parameter the researcher is interested in estimating? What does the parameter represent in this problem?

    **b.** Find *and* interpret the margin of error for estimating the parameter of interest with 90% confidence.

    **c.** Find *and* interpret a 90% confidence interval for the parameter of interest.

**25.** In a survey of 1219 US adults, 354 said that their favorite sport to watch on TV is women's college basketball.

    **a.** Estimate with 99% confidence the proportion of US adults who say that their favorite sport to watch on TV is women's college basketball.

    **b.** Using the answer you found in part a, can you be 99% confident that more than 25% of US adults' favorite sport to watch on TV is women's college basketball? Explain your answer.

    **c.** ESPN's CEO wants to determine the true proportion of US adults whose favorite sport to watch on TV is women's college basketball to be within 1% accuracy and with 95% confidence. What sample size should they take?

**26.** The Little League Association is interested in determining the average distance a male T-ball player can throw a T-ball. The data below represents the distance thrown in yards for twenty randomly selected male T-ball players.

| 12 | 21 | 20 | 23 | 17 | 12 | 15 | 16 | 14 | 10 |
|----|----|----|----|----|----|----|----|----|----|
| 13 | 23 | 12 | 8 | 18 | 17 | 20 | 19 | 6 | 18 |

    **a.** Estimate with 95% confidence the average distance a male T-ball player can throw a T-ball.

    **b.** The Little League Association has claimed in the past that male T-ball players can throw a T-ball farther than fifteen yards. At 95% confidence, is their claim correct?

# 6 HYPOTHESIS TESTING
## FOR ONE PARAMETER

*Statistical Inference* – the process of making judgments about a population based on properties of a sample randomly drawn from the population.

*Estimation* – involves approximating the value of an unknown parameter.

*Hypothesis Testing* – involves choosing between two opposing statements concerning a parameter.

## ➡ Section 1: What Is Hypothesis Testing? An Example to Walk Through:

According to M&M Mars Inc., the proportion of blue M&M's in their snack size pack is 10%. Suppose Ana really loves blue M&M's and always counts how many blue M&M's there are in every snack pack of M&M's she eats. (Ana's mom rewards her daughter's good behavior by giving her a snack pack of M&M's. Since Ana is a good girl, she eats many M&M's.) After many packages of M&M's, Ana begins complaining that there are not enough blue M&M's. So upon investigating this and keeping track of the number of blue M&M's, Ana and her mom discover that out of 240 M&M's, only sixteen were blue. At the 5% significance level can you conclude that the proportion of blue M&M's is less than 10%?

- What is the parameter of interest?

  Answer:

- What does M&M Mars Inc. claim to be true about this parameter?

  Answer:

- What does Ana think is true about this parameter?

  Answer:

- If I want to make a decision about this parameter _____, what would I calculate?

  Answer:

- In a hypothesis test, what we are trying to decide is, if the true value of the parameter, _____, is really equal to _____, is it likely or unlikely for us to calculate a value of _____ = _____.

- What can we look at to tell us if this value calculated, _____ is a likely or unlikely value if _____ is really equal to _____?

  Answer:

- What is the _____ of _____?

  Answer:

- What must be true in order for this to be the _____ of _____?

  Answer:

- If _____ is really _____, which value, A or B, labeled below is most likely to be obtained? Why?

  Answer:

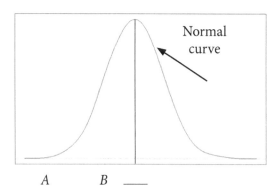

- What do we need to calculate to know where on this curve our _____ value falls?

  Answer:

- Is that value a surprising value?

  Answer:


- There are two ways for us to determine if this is a surprising value:

  **1.** Find the cut-off point such that any value in that range is a "surprising" value: These are the values that support Ana's claim.

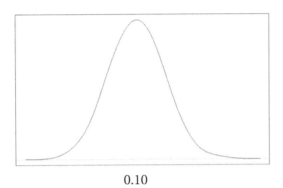

0.10

- How do we interpret the cut-off point?


  This says that any _____ value that results in a z-score _____ to _____ is a _____ value that shows support for Ana's claim.


  Our _____ resulted in the the z-score _____, so our _____ value _____ Ana's claim.


  **2.** Find the _____ of getting a _____ when _____ is really _____.

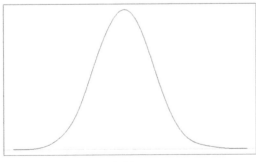

0.10

- How do we interpret this _____?

This implies that there is a _____ _____ of observing a

_____ if _____ is really _____.

Is this a small _____?

_____, since less than _____.

- Is _____, a likely or unlikely value if _____ is really_____?

Answer:

- Thus, what does this "evidence" suggest about the value of our parameter? Is there "enough" evidence to support Ana's claim?

Answer:

Let's now put a "definition name" to each step above. Then we will formally define the "steps" below.

# ➡ Section 2: Hypothesis Testing:

## Definitions

*Hypothesis*: a statement about one or more populations. Usually concerned about the value of a parameter.

*Research hypothesis*: the conjecture that motivates the research.

*Statistical hypothesis*: hypotheses that are stated in such a way that they may be evaluated by appropriate statistical techniques.

- The *null hypothesis* – Ho
  - the hypothesis being tested
  - we always assume throughout the test that the null hypothesis is true
  - a statement of agreement with conditions presumed to be true in the population
  - we either reject Ho or fail to reject Ho. **We never accept Ho!**
- The *alternative hypothesis* – Ha
  - corresponds to what the researcher is trying to prove

- <u>Types of Hypothesis</u>:

| *Two – Tailed* | *Right – Tailed* | *Left – Tailed* |
|---|---|---|
| Ho: $\mu = 50$ | Ho: $\mu = 50$ | Ho: $\mu = 50$ |
| Ha: $\mu \neq 50$ | Ha: $\mu > 50$ | Ha: $\mu < 50$ |

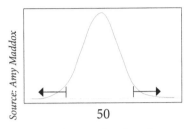

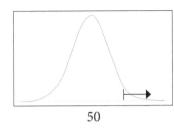

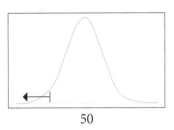

values in both tails
support Ha.

values in the right tail
support Ha.

values in the left tail
support Ha.

*Source: Amy Maddox*

*Level of Significance – α:* The probability that the test statistic will fall in the rejection region (also called critical region) if the null hypothesis is true is the *level of significance* of the test, which we denote as α. This is set by the researcher. Another way to think about α: It is how the researcher defines "a small probability." The values of α are usually 1%, 5%, or 10%. If the value is not stated we assume α equals 5%.

*Test Statistic:* The test statistic is the formula used to find the observed value, OV, of the test statistic, where the OV is a quantity that is used to make a decision in a test of hypothesis. The OV tells us where on the distribution curve the point estimate falls.

- General Form of the Test Statistic:

$$\text{T.S.} = \frac{\text{relevant statistic – hypothesized value}}{\text{standard error}}$$

*Assumptions of the Test Statistic*: The *assumptions of the test statistic* are properties that must be satisfied in order for your test statistic to be valid. The assumptions are what need to be true so that your test statistic will have the assumed distribution.

*Rejection Region* (RR): The *rejection region* (also called critical region) consists of those values of the test statistic that provide strong evidence in favor of the alternative hypothesis.

Recall, the RR describes for us the "region" on the distribution curve that "unlikely" values of the point estimate will fall if the assumed value of the parameter is true. Thus:

To find rejection region, if Ha:

| parameter $< a$ or | parameter $> a$ or | parameter $\neq a$ |
|---|---|---|
| RR: $Z \leq -z_o$ | $Z \geq z_o$ | $Z \leq -z_o$ or $Z \geq z_o$ |

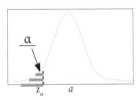

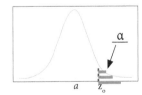

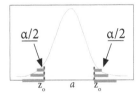

*If the OV falls in the RR, we reject Ho. Otherwise, we fail to reject Ho.*

*P-value*: The *p-value* is the probability of getting a value of the point estimate that is favorable or more favorable to the alternative hypothesis, if the null hypothesis is true.

The p-value describes for us, using probability, if the point estimate is an "unlikely" or "likely" value, if the assumed value of the parameter is true.

To find p-value, if Ha:

|  | parameter $< a$ | parameter $\neq a$ | parameter $> a$ |
|---|---|---|---|
| p-value: | (for ex:) $\Pr(\hat{p} < \underline{\quad}) = ?$ | $2\Pr(\hat{p} < \underline{\quad}) = ?$ <br> (if OV is negative) <br> $2\Pr(\hat{p} > \underline{\quad}) = ?$ <br> (if OV is positive) | $\Pr(\hat{p} > \underline{\quad}) = ?$ |

Small probabilities imply "unlikely," where "small" is defined as "less than α." Thus:

*If the p-value less than α, we reject Ho. Otherwise we fail to reject Ho.*

*Decision of the test*: The *decision* of the hypothesis test is either reject Ho or fail to reject Ho.

To determine the decision of the test, we ask ourselves two questions:

1. Is the OV in the RR? If we say yes, we reject Ho. If we say no, we fail to reject Ho.

2. Is the p-value $< α$? If we say yes, we reject Ho. If we say no, we fail to reject Ho.

There are many "steps" involved in hypothesis testing. In order to have your thoughts and work organized so that you can understand and complete hypothesis testing, you will follow the following steps when working a hypothesis problem. The steps are geared to walk you through the thought process of hypothesis testing.

- Steps for hypothesis testing:

1. Define population and parameter of interest.

2. State α = the level of significance.

3. Set up hypothesis:  Ho : parameter $= a$  versus  Ha: parameter $\neq a$

   or parameter $< a$

   or parameter $> a$

4. State the test statistic and give its distribution. (The distribution comes from the sampling distribution of the statistic.)

5. State and verify assumptions.

6. Analyze the data. In other words, state the point estimate and find the observed value of the test statistic.

7. How will you decide? In other words, find the rejection region and the p-value.

8. Conclusion – State your decision and your conclusion in terms of the problem.

**9.** If Ho is rejected, find and interpret the appropriate Confidence Interval.

To find the CI, if Ha:

| parameter $< a$ | parameter $\neq a$ | parameter $> a$ |
|---|---|---|
| pt.est. + \|RF\|*(SE) (using α) | pt.est. ± \|RF\|*(SE) (using α/2) | pt.est. – \|RF\|*(SE) (using α) |

# ➡ Section 3: Hypothesis Test for One Population Proportion, *p*:

For the population proportion,

Parameter:

Point Estimate:

Test Statistic:

Assumption:

Example 1: Consider the case of a software company that developed a general theme for a new game based on the assumption that half of its customers are over twenty years of age. The agency wants to change the general theme of this new game if this proportion has decreased. The agency conducted a survey of a sample of 400 of its customers. Of the 400 customers, 190 were over the age of twenty and 210 were twenty years or less. At the 5% significance level, what conclusion can be reached?

**1.** Population:                          Parameter of interest:

**2.** α =

**3.** Set up hypothesis:  Ho:

Ha:

**4.** What is your analysis based upon? (I.e., state the test statistic and give its distribution.)

**5.** State and check assumptions.

**6.** Analyze data.

**7.** How will you decide? (i.e., find the rejection region and the p-value)
   rejection region:

   p-value:

**8.** Decision:
   Conclusion:

**9.** If Ho is rejected, find and interpret the appropriate Confidence Interval.

Example 2: Pediatric researchers sampled 680 infants and found that 68% of them have had the MMR vaccine. Can we conclude on the basis of these data that the percent of all infants that have had the MMR vaccine is different than 62%? Assume that the researcher has set the level of significance to be 2.5%.

# Check Your Understanding

The Southern Online Viewing (SOW) Company claims that 60% of online video subscribers recommend its viewing platform over that of the Northern Online Viewing (NOW) Company's platform. To test this claim against the alternative that the actual proportion is less than 60%, a random sample of 50 online viewers was taken and 52% said they prefer SOW's platform over NOW's. At the 5% significance level, what conclusion can be reached?

# ➡ Section 4: Type of Errors and Power

Types of Errors: Every time we make a decision (either reject Ho or fail to reject Ho), we either made the correct decision or we made an error. There are two types of errors:

- *Type-I error*: The *Type-I error* is rejecting a true null hypothesis. We denote the probability of making a Type-I error as $\alpha$.

- *Type-II error*: The *Type-II error* is failing to reject a false null hypothesis. We denote the probability of making a Type-II error as $\beta$.

- *Power*: The *power* of the test is the probability of rejecting a false null hypothesis.

  Power = $1 - \beta$.

- Example: (Court of Law Example)

  Ho: innocent

  Ha: guilty

|  | | **Reality** | |
|---|---|---|---|
| | | Innocent<br>*Ho is true* | Guilty<br>*Ho is false* |
| **Jury Says** | Innocent<br>*Ho is true*<br>(fail to reject Ho) | Correct Action | Type II Error<br>Probability = $\beta$ |
| | Guilty<br>*Ho is false*<br>(reject Ho) | Type I Error<br>Probability = $\alpha$ | Correct Action |

- If we reject Ho, then we possibly made what type of error?

  Answer:

- If we fail to reject Ho, then we possibly made what type of error?

  Answer:

- Why is it that we never accept the Ho:

  Answer:

- Why is it that we do reject Ho:

  Answer:

- When choosing the test statistic, we want to choose the test statistic that has what two properties?

  **1.**

  **2.**

- What are three ways to maximize power:

  **1.**

  **2.**

  **3.**

- Suppose you have the following hypothesis:      Ho: House is on fire

  Ha: House is not on fire

  In the context of the null and alternative hypothesis, what is the Type-1 error?

  Answer:

In the context of the null and alternative hypothesis, what is the Type-2 error?

Answer:

In terms of the consequences of making the error, which of the above is the "worse" error?

Answer:

In hypothesis testing, when constructing the null and alternative hypothesis, they should be constructed in such a way that the "worse" error corresponds to a Type-_____ error.

Why?

# Check Your Understanding

1. Suppose you have the following null and alternative hypothesis:   Ho: It is too cold to ski
                                                                    Ha: It is not too cold to ski

    a. In the context of this problem, describe the power of the test.

    b. In the context of this problem, describe what $\alpha$ represents.

    c. In the context of this problem, describe what $\beta$ represents.

2. To increase power, should $n$ be increased or decreased?

3. To increase power, should $\alpha$ be increased or decreased?

4. Suppose the critical value of a two-tailed test equals $|2.064|$.

    a. What is the rejection region?

    b. Suppose the observed value equals $-3.45$. What is the decision of the test?

5. Suppose the p-value equals 0.0447 and the level of significance equals 5%. What is the decision of the test?

# Section 5: Hypothesis Test for μ When σ Is Unknown:

Recall the general form of the test statistic:

$$\text{T.S.} = \frac{\text{relevant statistic} - \text{hypothesized value}}{\text{standard error}}$$

Thus, for the population mean, the test statistic is:

Answer:

But once again, is it likely that we know σ?      So, we have for our formula:

Answer:

Thus we have for testing one population mean with the population standard deviation unknown:

Parameter:

Point Estimate:

Test Statistic:

Assumption:

**Example 1:** A manufacturer of motorcycle exhaust systems wants to test its new exhaust system to determine whether it meets state air pollution standards. The mean emission of all motorcycle exhaust systems of this type must be less than twenty parts per million of carbon. Fourteen exhaust systems were randomly selected and tested. The mean emission level was found to be seventeen ppm with standard deviation 2.94. Do the data supply sufficient evidence to allow the manufacturer to conclude that the new exhaust system for motorcycles meets the pollution standard? Assume that the manufacturer has set the probability of making a Type-I error to be to 1%. Show all steps.

If the data is given, we can let JMP do the work for us. Below is how the problem will look if data is given. The only steps that change are steps 5, 6, part of 7, and if needed step 9.

Example 2: A manufacturer of motorcycle exhaust systems wants to test its new exhaust system to determine whether it meets state air pollution standards. The mean emission of all motorcycle exhaust systems of this type must be less than twenty parts per million of carbon. Fourteen exhaust systems are manufactured for testing purposes. The data below represents the emission level of the fourteen exhaust systems tested. Do the data supply sufficient evidence to allow the manufacturer to conclude that the new exhaust system for motorcycles meets the pollution standard? Assume that the manufacturer has set the level of significance to 1%.

| 17 | 16 | 20 | 18 | 16 | 19 | 22 |
|----|----|----|----|----|----|----|
| 13 | 15 | 17 | 21 | 14 | 18 | 12 |

Show which steps change:

- JMP Instructions:
  1. Input data
  2. Under analyze, choose *distribution.* Click on the column containing the data, click on *Y, Columns* and click on *Ok.*
  3. Under the red options button next to distribution, choose *stack*
  4. Under red options button next to column 1 (or whatever you named your column), choose *normal quantile plot* (to check normality).
  5. Under red options button next to column 1 (or whatever you named your column), choose *test mean.*
  6. Type in the hypothesized value of the mean, hit okay.

- JMP Output:

**Distributions**

**Column 1**

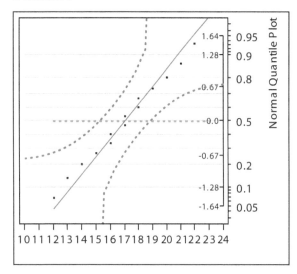

| Test Mean | |
|---|---|
| Hypothesized Value | 20 |
| Actual Estimate | 17 |
| DF | 13 |
| Std Dev | 2.9352 |

| | t Test |
|---|---|
| **Test Statistic** | -3.8243 |
| Prob > ltl | 0.0021* |
| Prob > t | 0.9989 |
| Prob < t | 0.0011* |

| One-sided Confidence Interval | | | | |
|---|---|---|---|---|
| **Parameter** | **Estimate** | **Lower CI** | **Upper CI** | **1-Alpha** |
| Mean | 17 | . | 19.07907 | 0.990 |
| Std Dev | 2.935198 | . | 5.222171 | 0.990 |

# Check Your Understanding

1. The Southern Online Viewing (SOW) Company would like to see if they are increasing their average length of viewing per customer over last year. The average amount of SOW content watched by a subscriber per day last year was 1.8 hours. A random sample of 68 viewers found that the average length of viewership was 1.85 hours with a standard deviation of 0.45 hours. At the 5% significance level, is there significant evidence to conclude the claim from SOW? Assuming all necessary assumptions are valid.

**2.** Suppose a researcher in public health is interested in determining if the average starting salary for those with a master's degree in public health is changing. The data below represents the starting salaries for 22 recent graduates that received a master's degree in public health. At the 1% significance level, can the researcher conclude that the average starting salary for those with a master's degree in public health is different from $55,000?

66.251  67.441   53.159   56.017   61.264   64.281   61.488   63.665   66.792   63.107   53.636

55.236  54.651   61.213   58.26    55.28    57.421   55.102   44.188   49.53    45.767   60.564

# Section 6: The Relationship between Confidence Intervals and Hypothesis Testing:

Recall when finding a CI, it was referred to as a $(1 - \alpha)$% CI. Thus for a 90% CI, $\alpha = 10$%.

Where else have we seen $\alpha$? It is the probability of making a Type-_____ error. Hence, there is a relationship between CI and hypothesis testing. A CI can be used to make a decision in a hypothesis test.

Recall that when we discussed CIs, in order for a CI to support a claim, _____ value in the CI has to support the claim.

Thus, to use a CI to draw a conclusion in a hypothesis test, we look to see if every value in the CI supports the alternative hypothesis. Another way to think of it is: if the null hypothesis could be true, we cannot reject it. Thus, if the value of the null hypothesis is contained within the CI, we must fail to reject Ho.

- Example 1: What percent of college students buy a pet? This is a question of concern for the local humane society. The local humane society would like to estimate with 95% confidence the proportion of college students that buy a pet. A random sample of 200 college students was obtained. Of those sampled, 44% had purchased a pet.

  - A 95% CI for _____:




  - The humane society would like to know if the true proportion of college students that purchase a pet is different than 50%. This gives us the following null and alternative hypothesis:

    Ho:

    Ha:

- Based on the CI, what is the decision and conclusion of the hypothesis test?

## Check Your Understanding

An educational researcher wants to determine if the percentage of high school students who like math is different from 50%. She takes a random sample of 85 high school students and finds that 32 of them like math.

  **a.** What is the null and alternative hypothesis that corresponds to this problem?

  **b.** Find a 95% confidence interval for the proportion of high school students who dislike math.

  **c.** Based on the confidence interval found in part b, what would be the decision of the hypothesis test for part a?

## ➡️ Section 7: Hypothesis Testing for One Parameter Overview:

Fill in the following for testing one population proportion:

Parameter:


Point Estimate:


General form for Ho:



Ha:



Test Statistic and its distribution:

Assumptions:

General form of CI:

Fill in the following for testing one population mean:

Parameter:

Point Estimate:

General form for Ho:

Ha:

Test Statistic and its distribution:

Assumptions:

General form of CI:

General form of the RR (for both parameters):

General form of probability statement for the p-value (for both parameters):

# Chapter 6 Homework:

## Section 3

Show ALL your work!

1. The La Leche Society has released an educational video on the benefits of breast feeding. Prior to the release of the videos, it was known that 75% of new moms breast-feed exclusively for the first six weeks of their child's life. The CEO wants to know if the videos are increasing this proportion. A random sample of 400 new moms was obtained and 320 breast-fed their new babies exclusively for the first six weeks. At the 5% significance level, what conclusion can be reached?

2. Although many new moms will nurse their new babies in the beginning, the Pediatrics Association believes that 50% of moms are still nursing their babies by the time the babies are six months old. The La Leche Society believes that this proportion is less than 50% and thus believes more educational information should be given to new moms to encourage the mothers to nurse for the first year of their child's life. A random sample of 425 new moms was obtained and 200 were still nursing their child at six months of age. At the 2.5% significance level, what conclusion can be reached?

3. With the increase in childhood obesity, the Pediatrics Association is concerned about the amount of physical activity children obtain daily. In 2012 the Pediatrics Association found that only 48% of children between the ages of five and twelve were physically active for at least one hour a day, where physically active refers to any activity that increases the child's heart rate to the aerobic zone. A random sample of 250 children between the ages of five and twelve was obtained. Of those sampled, 59% said they were physically active for at least one hour a day. At the 1% significance level, can the Pediatrics Association conclude that the proportion of children between the ages of five and twelve that are physically active for at least one hour a day has changed.

4. Many people assume that nurses are less likely than others to "feel faint" or "get sick" at the sight of blood. Suppose it is known that 30% of Americans will feel faint or get sick at the sight of blood. A random sample of 500 nurses was obtained. Of the 500 nurses, only 125 admitted to ever feeling faint or getting sick at the sight of blood. At the 10% significance level, can you conclude that nurses are less likely than other Americans to feel faint or get sick at the sight of blood?

# Section 4

**5.** Suppose you have the following null and alternative hypothesis:  Ho:  It is snowing

Ha:  It is not snowing

    **a.** In the context of this problem, describe a Type-I error.

    **b.** In the context of this problem, describe a Type-II error.

    **c.** In the context of this problem, describe the power of the test

**6.** Suppose the observed value of the test equals 2.45 and the critical value of the rejection region for a right tailed test equals 2.281. What is the decision of the test?

**7.** Suppose the observed value of the test equals -1.95 and the critical value of the rejection region for a left tailed test equals -2.371. What is the decision of the test?

**8.** Suppose the p-value for a two-tailed test equals 0.0463 and the level of significance equals 2.5%. What is the decision of the test?

**9.** Suppose the p-value for a two-tailed test equals 0.0241 and the level of significance equals 2.5%. What is the decision of the test?

**10.** What is the effect on $\alpha$ for the following scenarios:

    **a.** $\beta$ is increased

    **b.** $\beta$ is decreased

    **c.** power is increased

    **d.** power is decreased

11. What is the effect on the power of the test for the following scenarios:

   **a.** $\beta$ is increased

   **b.** $\beta$ is decreased

   **c.** $\alpha$ is increased

   **d.** $\alpha$ is decreased

   **e.** $n$ is increased

   **f.** $n$ is decreased

12. What is the effect on $\beta$ for the following scenarios:

   **a.** $n$ is increased

   **b.** $n$ is decreased

   **c.** $\alpha$ is increased

   **d.** $\alpha$ is decreased

13. Suppose your roommate believes the mean amount it will cost to repair your dorm refrigerator is more than \$50. To determine if she is correct, suppose you perform a hypothesis test with level of significance equal to 5%. For each choice below, determine if you made the correct decision or if you made an error. If an error was made, determine if a Type-I or Type-II error was committed.

   **a.** You conclude that the mean amount to repair a dorm refrigerator is more than \$50 when in reality the mean amount to repair a dorm refrigerator is not more than \$50.

   **b.** You conclude that the mean amount to repair a dorm refrigerator is not more than \$50 when in reality the mean amount to repair a dorm refrigerator is not more than \$50.

   **c.** You conclude that the mean amount to repair a dorm refrigerator is not more than \$50 when in reality the mean amount to repair a dorm refrigerator is more than \$50.

   **d.** You conclude that the mean amount to repair a dorm refrigerator is more than \$50 when in reality the mean amount to repair a dorm refrigerator is more than \$50.

14. For the each of the null and alternative hypothesis below, determine if it represents a valid null and alternative hypothesis. If the null or alternative hypothesis is not valid, explain why.

a.  Ho: $\mu = 20$
    Ha: $\mu \geq 20$

b.  Ho: $\mu \neq 18$
    Ha: $\mu = 18$

c.  Ho: $\mu = 0$
    Ha: $\mu < 0$

d.  Ho: $\bar{x} = 15$
    Ha: $\bar{x} > 15$

e.  Ho: $\mu = 20$
    Ha: $\mu > 19$

f.  Ho: $p = 5$
    Ha: $p > 5$

g.  Ho: $p < 0.20$
    Ha: $p > 0.20$

h.  Ho: $p = \frac{1}{2}$
    Ha: $p \neq \frac{1}{2}$

# Section 5

Show all work! For the problems containing data, you must use JMP to analyze the data.

15. Are college students spending more or less time studying for class than in the past? This was a question of interest for researchers. Twenty years ago, college students spent an average of 2.8 hours a day studying for class. A random sample of fifty present day college students was obtained and the average time they spend studying for class was found to be 2.4 hours with standard deviation 1.4. At the 2% significance level can the researchers conclude that the average amount of time college students spend studying for class is different than the average amount of time in the past spent studying? Assume the time spent studying is a normally distributed random variable.

16. Is the strep virus getting harder or easier to heal? Pediatricians are interested in studying the average length of time patients with strep are running fever while on antibiotics. In previous years, the average length of time patients ran fever with strep while on antibiotics was 1.8 days. A random sample of forty-six recent patients with strep ran fever while on antibiotics for an average of 2.3 days with standard deviation 1.2. At the 5% significance level, can pediatricians conclude that the average amount of time patients with strep run fever while on antibiotics has changed? Assume the random variable has a normal distribution.

17. Dieticians claim that most adults do not intake enough dietary fiber from food (not supplements). For optimal health, adults should intake an average of twenty-eight grams of fiber from food per day. Below is the amount of dietary fiber from food consumed by twenty randomly selected adults. At the 5% significance level, can we conclude that the average amount of dietary fiber from food consumed per day is less than twenty-eight grams?

| 22 | 23 | 18 | 15 | 17 | 14 | 25 | 20 | 21 | 29 |
|----|----|----|----|----|----|----|----|----|----|
| 28 | 14 | 12 | 15 | 18 | 13 | 17 | 21 | 24 | 19 |

**18.** Is the flu making us sicker? A question of interest at the latest health summit was whether the latest strain of flu was causing adults to run a higher fever than previous strains. The average temperature for previous strains of flu was 101.4 degrees. A random sample of eighteen adults with the latest strain of flu was obtained. The data below represents their highest temperature while sick with the latest flu strain. Based on the data, can we conclude that the latest strain of flu is causing patients to run a higher fever, on average, than previous strains of flu? Let the probability of making a Type-I error be 10%.

102.1   101.8   100.9   103.8   102.8   100.8   102.5   102.0   103.1

104.0   104.2   101.1   100.3   101.9   101.7   101.8   102.3   104.1

# Section 6

**19.** With economical concerns being so great, many people are no longer purchasing new vehicles, but are instead keeping their vehicles for a longer period of time, until they no longer work. The American Automobile Association is interested in determining if the average age of vehicles for Americans is different than six years. A random sample of 35 Americans was obtained. The average age of their vehicles was 6.2 years with standard deviation 1.2.

**a.** Set up the null and alternative hypothesis:

**b.** At the 1% significance level, use a CI to determine if the average age of vehicles for Americans is different than six years. Make sure and give your CI and your decision and conclusion.

**20.** Suppose a 95% CI for the proportion of adults that regularly attend a tae kwon do class at least once a week was found to be (0.2534, 0.3894).

**a.** For the null and alternative hypothesis: Ho: $p = 0.40$ versus Ha: $p \neq 0.40$, what is your decision for this test of hypothesis?

**b.** Based on the CI, which of the following statements gives the best conclusion about the true value of the proportion of adults that regularly attend a tae kwon do class at least once a week?

**i.** I am 95% confident that the proportion of adults that regularly attend a tae kwon do class at least once a week is greater than 40%.

**ii.** I am 95% confident that the proportion of adults that regularly attend a tae kwon do class at least once a week is different than 40%

**iii.** I am 95% confident that the proportion of adults that regularly attend a tae kwon do class at least once a week is less than 40%

**iv.** I am 95% confident that the proportion of adults that regularly attend a tae kwon do class at least once a week equals 40%

# Combining the Sections

**21.** In a random sample of 28 four-person tents, the mean price was $344.19 and the standard deviation was $61.32. Assume the prices are normally distributed. At the 5% significance level can we conclude that the mean price of four-person tents has increased from the previous mean price of $325?

**22.** A hot topic of debate is whether the legal driving age should be raised to eighteen. From a random sample of 400 Americans, 52% favored raising the legal driving age to eighteen. Based on the data, can you conclude that a majority of Americans believe the legal driving age should be raised to 18?

Let $\alpha = 2.5\%$

**23.** Are our kids getting taller? The average height of kindergartners in 1950 was 39.5 in. The data below represents the height (in inches) of twenty-six randomly selected kindergartners. At the 5% significance level, what conclusion can be reached?

| 40.1 | 39.9 | 41.2 | 41.3 | 39.2 | 39.3 | 38.7 | 38.8 | 39.2 | 40.3 | 41.1 | 40.7 | 39.8 |
| 39.7 | 40.9 | 40.0 | 39.0 | 41.0 | 40.6 | 39.6 | 39.5 | 40.5 | 40.8 | 40.9 | 39.2 | 41.1 |

**24.** Due to the economy, are Americans staying in their first professional job longer? This was a question of interest for the Federal Jobs Bureau. In 2010, Americans were staying an average of 2.3 years in their first professional job. The data listed below is the length of time (in years) that eighteen randomly selected adults were staying in their first professional job. At the 5% significance level, what conclusion can be reached?

| 2.4 | 2.1 | 1.9 | 3.1 | 3.2 | 2.9 | 2.6 | 1.9 | 3.8 |
| 2.3 | 2.5 | 1.8 | 2.2 | 4.5 | 5.1 | 2.1 | 2.4 | 6.1 |

# HYPOTHESIS TESTING FOR TWO PARAMETERS

CHAPTER 7

Many times we are interested in comparing two populations, thus we are interested in comparing two parameters. For CI and for hypothesis testing, the concepts and general idea are the same. So, to determine the formula for the CI or to determine which hypothesis test to use, we just need to know based on the parameter of interest, what is the point estimate and what is the distribution and standard deviation of the point estimate.

## ➡ Section 1: Independent vs. Dependent Samples:

When testing *two* population parameters, one criterion in choosing the test statistic depends on whether the samples are independent or dependent. Thus, the first question we should ask ourselves when we realize we are testing two parameters is "Are the samples independent or dependent?" In other words, "Did we obtain the samples in such a way that they are independent or did we obtain them in such a way that they are dependent?"

- Two samples are *independent* if the data values obtained from one sample are unrelated to the values obtained in the second sample.

- The samples are *dependent* if each data value from one sample is paired in a natural way with a data value from another sample.

- Why use "paired" or "dependent" samples? Because it can eliminate extraneous or confounding factors.

  For Example: Suppose we want to test two sunscreens—Sunscreen A and Sunscreen B.

- Experiment 1: We could randomly assign one group to receive Sunscreen A and randomly assign another group to receive Sunscreen B. We then look at the amount of sun damage after a specific amount of time.

  - Is this an example of independent or dependent samples?
    Answer:


  - Is there something other than the brand of sunscreen that can affect the amount of sun damage?
    Answer:

- Experiment 2: We could randomly choose one group of people and apply Sunscreen A to the right half of their back and Sunscreen B to the left half of their back. We then look at the amount of sun damage after a specific amount of time.

  - Is this an example of independent or dependent samples?
  Answer:

  - Does this control for the extraneous factor?
  Answer:

- How do we obtain "paired" or "dependent" samples?

  1. have pre and post measurements from the same subject

  2. have litter mates of same sex

  3. twins

  4. match people based on some characteristic

# ➡ Section 2: Hypothesis Testing for Two Population Means with Dependent Samples:

When testing two population means from dependent samples we analyze the *differenced data*.

- For example, suppose we have the following data set from two dependent samples:

| $X_1$ | $X_2$ | $X_d = X_1 - X_2$ |
|:-----:|:-----:|:-----------------:|
| 66 | 60 | 6 |
| 82 | 79 | 3 |
| 96 | 92 | 4 |
| ⋮ | ⋮ | ⋮ |

- We define a new random variable, $X_d = X_1 - X_2$ and a new parameter $\mu_d = \mu_1 - \mu_2$. This takes us from two samples to one sample and hence from testing two parameters to testing one parameter: the population differenced mean, $\mu_d$. So, to proceed, we do as before: if $\sigma$ is known we perform the $z$-test: if $\sigma$ is unknown we perform the $t$-test.

- Thus, the parameter of interest:

- The point estimate:

- Null and Alternative hypothesis:

  - If we want to determine if $\mu_1 \neq \mu_2$, then this is equivalent to testing

    Ho:                          Ha:

- If we want to determine if $\mu_1 > \mu_2$, then this is equivalent to testing

    Ho:                              Ha:

- If we want to determine if $\mu_1 < \mu_2$, then this is equivalent to testing

    Ho:                              Ha:

- Test Statistic and its distribution:

    Answer:


- General form of the CI:

    Answer:


- Assumption:

    Answer:


- Example 1: An educational specialist for kindergarten students was interested in finding out whether a creative approach program would increase the scores of kindergartners on a reading ability test. Sixteen students were selected and paired according to a previous test on letter recognition ability. The scores on the reading ability test are as follows:

| Creative Approach Program | Standard Program |
|---|---|
| 76 | 69 |
| 89 | 91 |
| 96 | 92 |
| 82 | 73 |
| 92 | 85 |
| 77 | 69 |
| 85 | 82 |
| 85 | 86 |

At the 5% significance level, do the data suggest that the creative approach program is effective in increasing the scores of the kindergartners on the reading ability test?

Does the CI agree with your decision in step 8? Why or Why not?

Answer:

JMP Instructions:

1. Put the data into two separate columns
2. Under *columns*, choose *new column*
3. Name the column *difference*
4. Choose *column properties*
5. Choose *formula*
6. Click on column 1
7. Click on minus sign
8. Click on column 2
9. Hit okay
10. Under *Analyze*, choose *distribution*
11. Click on difference column, click on *Y*, hit okay
12. Follow the steps for using a *t*-test

JMP Output:

### *Distributions*
### *d in test scores*

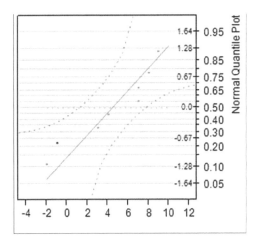

| Test Mean | |
|---|---|
| Hypothesized Value | 0 |
| Actual Estimate | 4.375 |
| DF | 7 |
| Std Dev | 4.13824 |
| | *t Test* |
| Test Statistic | 2.9903 |
| Prob > |t| | 0.0202* |
| Prob > t | 0.0101* |
| Prob < t | 0.9899 |

| One-sided Confidence Interval | | | | |
|---|---|---|---|---|
| Parameter | Estimate | Lower CI | Upper CI | 1-Alpha |
| Mean | 4.375 | 1.603066 | | 0.950 |
| Std Dev | 4.138236 | 2.919184 | | 0.950 |

# Check Your Understanding

An article written in Healthy Choices magazine claimed that you would gain weight once you subscribe to an online movie repository site. Data analysts for the Southern Online Viewing (SOW) Company wanted to prove the article wrong, so they surveyed 20 new subscribers and measured their weight. After two years they followed up with them to measure their weight again. The data is given below. At the 1% significance level, do the analysts have evidence to support the article's claim?

| Weight Before | 128 | 112 | 124 | 116 | 119 | 127 | 115 | 126 | 121 | 120 |
|---|---|---|---|---|---|---|---|---|---|---|
| Weight After | 129.5 | 118 | 130.5 | 118 | 128 | 133 | 120.5 | 131 | 120 | 118 |

| Weight Before | 119.5 | 128 | 115.5 | 121 | 119.5 | 117.5 | 120.5 | 118.5 | 111 | 122 |
|---|---|---|---|---|---|---|---|---|---|---|
| Weight After | 126 | 129.5 | 124 | 125 | 117 | 122 | 124.5 | 116.5 | 114.5 | 129 |

# Practical Significance – VS – Statistical Significance:

Based on the CI, do you believe that the Creative Approach program does a "good" job on raising the reading ability for kindergartners? What if we know the program is expensive to implement?

Answer:

As the superintendent of the school district looking to incorporate the program, knowing that it will cost quite a bit of money to implement the Creative Approach program, would you keep the alpha level set at 5% or would you lower the alpha level to say, 1%?

Answer:

As the superintendent, if you lower the alpha level to 1%, without "redoing" any work, what would your decision be?

Answer:

In interpreting a hypothesis test result, one should always take into account that a test can show a statistically significance difference and yet in reality it could be the case that there is not enough difference to be practically significant! This is why a good statistical report should give the p-value and the CI, since they help us in determining practical significance as well as statistical significance.

# ➡ Section 3: Hypothesis Testing for Two Population Means with Independent Samples:

When testing two population means from independent samples, we cannot analyze the differenced data like we did for dependent samples. We must keep the data sets separate and choose the appropriate test; the test that controls for alpha and maximizes power. Once again the main idea for hypothesis testing and CI remain the same. We just need to know the following:

- The parameter of interest:

- The point estimate:

- General form for Null and Alternative hypothesis:
  Answer:


- Test Statistic and its distribution:
  Answer:






- Assumptions:
  Answer:




- General form of the CI:
  Answer:

- Again, when using the pooled *t*-test, we have **two** assumptions that we have to check (in step 5):

  **1.** Verifying normality:

  **2.** Verifying equal population variances: We will verify this using JMP, where JMP is doing a hypothesis test for the null and alternative hypothesis.

  Ho:                      vs.                      Ha:

  - Do we want to        reject Ho      or      fail to reject Ho?

  - To make our decision, JMP will provide us a p-value. Do we want a "large"      or      "small"      p-value?

  - How do we define a _____ p-value?

  - Are we more concerned here with a Type-1 or Type-2 error?

    - What is a Type-1 error here?

    - What is a Type-2 error here?

    - Which is "worse"?

  - Thus, we will define a _____ p-value as p-value _____ .

- Example 1: Do students taking Introduction to Statistics in early morning classes perform better than their peers in late afternoon classes? The following are the student's final exam scores. Using a 5% level of significance, does the early morning class seem to perform better on the final exam than the late afternoon class?

| _**Early Morning Class**_: | 80 | 82 | 71 | 90 | 69 | 75 | 78 | 87 | |
|---|---|---|---|---|---|---|---|---|---|
| | 95 | 98 | 86 | 79 | 77 | 84 | 81 | 91 | |
| _**Late Afternoon Class**_: | 79 | 83 | 73 | 77 | 84 | 90 | 91 | 95 | 80 |
| | 94 | 75 | 76 | 69 | 89 | 80 | 82 | 81 | |

Do we need to perform step 9? Regardless, let's perform step 9:

Does the CI agree with your decision in step 8? Why or Why not?
Answer:

JMP Instructions:

1. Enter all data into one column, name the column *y*

2. Under *columns*, choose *new column*, name the column *x*

3. Under *data type*, choose *character*

4. Click *okay*

5. Label the data in the "*x*" column – 1's for the first data set, 2's for the second data set

6. Under *analyze*, choose *fit y by x*

7. Click on *y* column, click on *y*

8. Click on *x* column, click on *x*

9. Choose *normal quantile* plot

10. Choose *unequal variances*

11. Choose *means/anova/pooled t-test*

JMP Output:

### Oneway Analysis of final exam scores By x

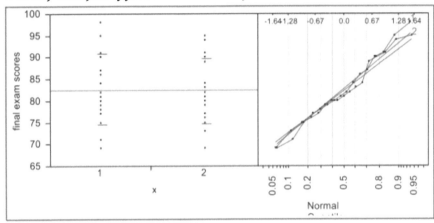

**Oneway Anova**

**t Test**

2-1

| *Assuming equal variances* | | | |
|---|---|---|---|
| Difference | -0.4522 | t Ratio | -0.16625 |
| Std Err Dif | 2.7201 | DF | 31 |
| Upper CL Dif | 5.0954 | Prob > ltl | 0.8690 |
| Lower CL Dif | -5.9998 | Prob > t | 0.5655 |
| Confidence | 0.95 | Prob < t | 0.4345 |

**Tests that the Variances are Equal**

| *Level* | *Count* | *Std Dev* | *MeanAbsDif to Mean* | *MeanAbsDif to Median* |
|---|---|---|---|---|
| 1 | 16 | 8.170832 | 6.523438 | 6.437500 |
| 2 | 17 | 7.454272 | 5.923875 | 5.882353 |

| *Test* | *F Ratio* | *DFNum* | *DFDen* | *p-Value* |
|---|---|---|---|---|
| O'Brien[.5] | 0.1934 | 1 | 31 | 0.6632 |
| Brown-Forsythe | 0.1143 | 1 | 31 | 0.7375 |
| Levene | 0.1498 | 1 | 31 | 0.7013 |
| Bartlett | 0.1264 | 1 | . | 0.7222 |
| F Test 2-sided | 1.2015 | 15 | 16 | 0.7181 |

# Check Your Understanding

Suppose you are interested in determining if men make more money than women, on average. To test, a random sample of ten men and ten women is obtained. At the 5% significance level, can you conclude that the average salary for men is higher than it is for women? The data given below represents the salaries in thousands.

| *Men* | 67.441 | 53.159 | 56.017 | 61.264 | 64.281 | 61.488 | 63.665 | 66.792 | 63.107 | 53.636 |
|---|---|---|---|---|---|---|---|---|---|---|
| *Women* | 54.651 | 61.213 | 58.26 | 55.28 | 57.421 | 55.102 | 44.188 | 49.53 | 45.767 | 60.564 |

# ⮕ Section 4: Hypothesis Testing for Two Population Proportions for Independent Samples:

When testing two population proportions, our first question should be what it is for testing two population means: "Are the samples independent or dependent?" Because of time, we will only consider the case where the samples are *independent*. As always, we want to choose the appropriate test; the test that controls for alpha and maximizes power. Recall, once again the main idea for hypothesis testing and CI remain the same. We just need to know the following:

- The parameter of interest = _____

- The point estimate = _____

- General form for Null and Alternative hypothesis:

  Answer:


- Test Statistic and its distribution:

$$Z_0 = \frac{(\hat{p}_1 - \hat{p}_2) - (p_1 - p_2)}{\sqrt{\bar{p} \cdot \bar{q} \left( \frac{1}{n_1} + \frac{1}{n_2} \right)}} \overset{.}{\sim} N(0,1)$$

$$\text{where } \bar{p} = \frac{n_1 \hat{p}_1 + n_2 \hat{p}_2}{n_1 + n_2}.$$

- General form of the CI:

$$(\hat{p}_1 - \hat{p}_2) \pm z_{a/2} \sqrt{\frac{\hat{p}_1 \cdot \hat{q}_1}{n_1} + \frac{\hat{p}_2 \hat{q}_2}{n_2}}$$

- Assumptions:

  $n_1{}^*p_1$ and $n_1{}^*q_1$ are both greater than 5 (for both hypothesis test and for CI, you have to use $\hat{p}$ and $\hat{q}$.) $n_2{}^*p_2$ and $n_2{}^*q_2$ are both greater than 5

- JMP does not use this test to analyze two proportions from independent samples. The test it uses is the WALD test, which for two proportions has an approximate normal distribution. When we use JMP, JMP will only give us a p-value and a CI. It does not give us a traditional observed value.

The p-value represents the probability of the populations having the characteristic of interest:

| *Adjusted Wald Test* |
| --- |
| P(BIPOLARINOT VCFS)-P(BIPOLARIVCFS) ≥ 0 |
| P(BIPOLARINOT VCFS)-P(BIPOLARIVCFS) ≤ 0 |
| P(BIPOLARINOT VCFS)-P(BIPOLARIVCFS) = 0 |

You choose the p-value that corresponds to the alternative hypothesis.

The CI is a CI on the difference of the two proportions. You have to be careful with JMP because JMP is going to put the characteristics in alphabetical order. So you want to look at the output to see if JMP calculated the CI on $p_1 - p_2$ or on $p_2 - p_1$.

**Two Sample Test for Proportions**

| Description | Proportion Difference | Lower 95% | Upper 95% |
|---|---|---|---|
| P(BIPOLARINOT VCFS)-P(BIPOLARIVCFS) | -0.09929 | -0.15663 | -0.0411 |

- Example: Velo-Cardio-Facial Syndrome (VCFS) is a syndrome characterized by having more than 180 clinical features both physical and behavioral including congenital heart disease, immune disorders, feeding problems, cleft palate, development disorders, and psychological disorders such as ADHD, bi-polar disorder, and schizophrenia. Researchers would like to know if the proportion of those with VCFS having bi-polar disorder is greater than the proportion of those without VCFS having bi-polar disorder. A random sample of 400 adults with VCFS was obtained and 114 of them were found to be bi-polar. A random sample of 420 adults without VCFS was obtained and seventy-eight of them were found to be bi-polar. At the 5% significance level, what conclusion can the researchers make?

    1.

    2.                         3.

    4.

    5. p-value:

    6. CI:

**7.** Decision: (Using p-value and CI)

**8.** Conclusion:

---

JMP Instructions:

**1.** Create three columns:

A *column* describing the two populations. (If you use numbers instead of letters you will have to make the column a "character" column. Ex: VCFS, not VCFS.)
A *column* describing the characteristic of interest. (If you use numbers instead of letters you will have to make the column a "character" column. Ex: Bipolar, not Bipolar.)
A *frequency* column. Enter the frequency data. Make sure your numbers match the pop/ char of interest. (This column should NOT be a character column.)

**2.** Under *distributions*, choose fit *y* by *x*.
Click on the "*population*" column, click on *x*.
Click on the "*characteristic of interest*" column, click on *y*.
Click on the "*frequency*" column, click on Freq.

**3.** Under the red options button, choose *two sample test for proportions*.

---

- JMP Output:

**Contingency Analysis of BIPOLAR By VCFS**

Freq: FREQUENCY

**Two Sample Test for Proportions**

| Description | Proportion Difference | Lower 95% | Upper 95% |
|---|---|---|---|
| P(BIPOLARINOT VCFS)-P(BIPOLARIVCFS) | -0.09929 | -0.15663 | -0.0411 |

| Adjusted Wald Test | Prob |
|---|---|
| P(BIPOLARINOT VCFS)-P(BIPOLARIVCFS) ≥ 0 | 0.9996 |
| P(BIPOLARINOT VCFS)-P(BIPOLARIVCFS) ≤ 0 | 0.0004* |
| P(BIPOLARINOT VCFS)-P(BIPOLARIVCFS) = 0 | 0.0008* |

# Check Your Understanding

Health researchers are interested in determining if more men smoke than women do. A random sample of 380 men included 97 who smoke, and a random sample of 345 women included 69 who smoke. Use a 0.1 significance level to test the claim that the proportion of men who smoke is greater than the proportion of women who smoke.

## ➡ Section 5: Hypothesis Testing for Two Parameters Overview:

Fill in the following for testing two population means from dependent samples:

When performing a hypothesis test for two population means from dependent samples, we analyze the

_____ data. Thus we have a new random variable:

Parameter:


Point Estimate:


General form for Ho:


Ha:


Test Statistic and its distribution:


Assumptions:


General form of CI:

Fill in the following for testing two population means from independent samples:

Parameter:

Point Estimate:

General form for Ho:

Ha:

Test Statistic and its distribution:

Assumptions:

General form of CI:

General form of the RR (for both methods):

General form of the probability statement of the p-value (for both methods):

Fill in the following for testing two population proportions from independent samples:

Parameter:

Point Estimate:

General form for Ho:

Ha:

Test Statistic and its distribution:

Assumptions:

The _____ Test just gives the p-value and the CI for the difference in the proportions. So,

If CI on $(p_1 - p_2) = (-a, -b)$, what is the decision of the test?

What does that imply about the relationship between $p_1$ and $p_2$?

If CI on $(p_1 - p_2) = (a, b)$, what is the decision of the test?

What does that imply about the relationship between $p_1$ and $p_2$?

If CI on $(p_1 - p_2) = (-a, b)$, what is the decision of the test?

What does that imply about the relationship between $p_1$ and $p_2$?

# ➡️ Chapter 7 Homework:

## Section 2

1. Does the bat make a difference? This was a question of interest at the regional little league monthly meeting. To test whether the more expensive bats do increase the average distance hit, ten little league batters of the same age were randomly chosen. The batters hit ten balls with the top of the line bat and after a rest; hit ten balls with the lower price bat. The data below represents the distance hit in feet for the batters.

| Top of the Line Bat | 69.9 | 59.2 | 58.9 | 58.7 | 63.2 | 62.5 | 66.5 | 70.4 | 65.9 | 55.5 |
|---|---|---|---|---|---|---|---|---|---|---|
| Lower Price Bat | 59.2 | 61.2 | 52.6 | 54.3 | 58.1 | 62.3 | 56.5 | 62.4 | 53 | 61.8 |

   a. At the 5% significance level, does the evidence suggest that the top of the line bat does increase the average distance hit versus the lower price bat?

   b. Do you think there is a confounding factor that was not controlled for in the design of the experiment?

2. Does owning an indoor dog really help bring down borderline high systolic blood pressure? To answer this question, a random sample of twelve subjects with chronically borderline high systolic blood pressure was obtained. Their systolic blood pressure was recorded. They then received a house-trained dog to live indoor with them. After one month, their blood pressure was taken again. At the 2.5% significance level, can you conclude that owning a pet will reduce the average systolic blood pressure?

| Blood Pressure Before | 139.9 | 130.4 | 131.3 | 130.8 | 128.9 | 129.5 | 130.8 | 129.3 | 128.1 | 133.4 | 125.2 | 123.9 |
|---|---|---|---|---|---|---|---|---|---|---|---|---|
| Blood Presure After | 136.3 | 127.4 | 125.6 | 132.2 | 127.5 | 126.9 | 131.1 | 123.7 | 130.4 | 134.8 | 125.6 | 124.9 |

3. A popular myth is that for identical twins, one is the "nice one" and one is the "mean one." Fourteen pairs of identical twins were randomly selected and given a personality test for measuring "niceness" or "meanness" where the higher score implies "niceness" and a lower score implies "meanness." At the 1% significance level, is there a difference in the personality scores between the identical twins.

| First Born Score | 43 | 54 | 49 | 53 | 49 | 52 | 58 | 51 | 59 | 48 | 45 | 49 | 47 | 46 |
|---|---|---|---|---|---|---|---|---|---|---|---|---|---|---|
| Second Born Score | 52 | 47 | 50 | 46 | 50 | 51 | 50 | 45 | 46 | 47 | 49 | 54 | 50 | 54 |

# Section 3

**4.** Do smokers have higher blood pressure? A random sample of twenty-eight smokers was obtained and their systolic blood pressure was obtained. Another random sample of thirty non-smokers was obtained and their systolic blood pressure was obtained.

| Smokers Blood Pressure | | | Nonsmokers Blood Pressure | | |
|---|---|---|---|---|---|
| 127 | 134 | 143 | 106 | 116 | 130 |
| 141 | 135 | 138 | 118 | 121 | 126 |
| 129 | 137 | 130 | 123 | 118 | 117 |
| 136 | 135 | 129 | 117 | 118 | 125 |
| 144 | 138 | 131 | 122 | 112 | 115 |
| 128 | 136 | 140 | 117 | 111 | 127 |
| 131 | 133 | 140 | 118 | 128 | 116 |
| 137 | 133 | 133 | 124 | 119 | 116 |
| 139 | 131 | 139 | 126 | 116 | 114 |
| 136 | | | 125 | 109 | 113 |

**a.** At the 5% significance level, what conclusion can be reached?

**b.** Is there a possible confounding factor that was not controlled for in the experiment?

**5.** The BMI levels for a random sample of thirty females with a college degree and for a random sample of thirty-two females with a high school diploma is listed in the table below. At the 5% significance level, can you conclude that the average BMI level for females with a college degree is different than the average BMI level for females with just a high school degree?

| BMI Level For H.S. Graduates | | | | BMI Level For College Graduates | | | |
|---|---|---|---|---|---|---|---|
| 19.1 | 21.3 | 21 | 17.3 | 21.6 | 19.4 | 21.2 | 23.2 |
| 19.8 | 20 | 19.2 | 17.8 | 22.6 | 18.8 | 20.6 | 21.1 |
| 19.8 | 18.3 | 18.1 | 18.8 | 20.4 | 20.8 | 21.2 | 18.4 |
| 19.1 | 19.2 | 20.6 | 17.4 | 21.4 | 21.5 | 22.6 | 20.1 |
| 19.5 | 17.3 | 19.5 | 17.2 | 21.2 | 21.3 | 20.8 | 21.2 |
| 19.6 | 18.6 | 18.5 | 18.5 | 21.6 | 20.7 | 20.5 | 22.2 |
| 17.9 | 19.3 | 18.8 | 18.4 | 21.4 | 22.6 | 21.4 | |
| 19.6 | 19.5 | 17.4 | 21.4 | 19.9 | 21.7 | 19.5 | |

**6.** Many children and even adults will not eat the bread crust. But is that the best part to eat in terms of the fiber content? It is well known that wheat bread has more fiber than white bread, but what amount of the fiber is contained in the crust? A study was done to determine if eating white bread with the crust is just as good or better for you in terms of fiber content as eating wheat bread without the crust. At the 2.5% significance level, can you conclude that there is a more fiber in the white bread with the crust than there is in wheat bread without the crust?

| White With Crust | 1.5 | 1.08 | 1.11 | 1.21 | 1.13 | 1.3 | 1.14 | 1.1 | 1.11 | 1.2 | |
| | 1.15 | 1.2 | 0.97 | 0.98 | 1.27 | 0.88 | 1.11 | 0.93 | 0.88 | 1.15 | |
| Wheat Without Crust | 1.13 | 1.16 | 0.73 | 1.03 | 0.81 | 1.17 | 0.67 | 1.19 | 1.15 | 1.4 | 1.04 |
| | 1.13 | 1.02 | 1.33 | 0.55 | 1.02 | 1.47 | 0.93 | 1.09 | 0.94 | 0.95 | 0.98 |

# Section 4

**7.** Feverfew is an herb that many claim will cure a migraine. Does it work better than standard migraine medicine? To answer this question, one hundred randomly selected migraine sufferers took feverfew at the onset of a migraine. Of the one hundred subjects, eighty-eight reported relief from the migraine. Another random sample of 125 migraine sufferers took the standard migraine medicine at the onset of a migraine. Of those 125 subjects, seventy-six reported relief from the migraine. At the 5% significance level, is there a difference in the proportion of subjects that reported relief from the migraine for the two types of cures: feverfew and standard migraine medicine?

**8.** Many claim that white noise such as a vacuum cleaner or large box fan will help soothe colicky babies. To test this claim, 130 randomly selected moms of colicky babies rocked their baby in a room with a large box fan turned on high. Of the 130 moms, ninety reported that the large box fan did indeed soothe their baby. Another random sample of 150 moms of colicky babies rocked their baby in a room without any white noise. Of the 150 moms, ninety-four reported that they were able to soothe their baby without using any white noise, simply by rocking them. At the 10% significance level, can you conclude that white noise does help soothe a colicky baby?

**9.** Many people think of fishing as a boy's activity. But is that true? Do boys like fishing more than girls? To answer this question, a random sample of 200 boys between the ages of eight and twelve was obtained. Of the 200 boys, 124 said they like to fish. Another random sample of 190 girls between the ages of eight and twelve was obtained. Of the 190 girls, fifty-four said they like to fish. At the 2.5% significance level, what conclusion can be reached?

# Combining the Sections

**10.** There is a theory that people who drink caffeine can run longer than those who do not. To test this theory a random sample of seven people who drink caffeine and a random sample of seven people who do not drink caffeine was obtained and their length of time they could run non-stop was measured, in seconds.

| Non-Caffeine drinkers | 359 | 280 | 138 | 227 | 203 | 184 | 231 |
|---|---|---|---|---|---|---|---|
| Caffeine drinkers | 394 | 477 | 439 | 428 | 391 | 488 | 454 |

   **a.** At the 5% significance level, can you conclude that those who drink caffeine can run longer than those who do not?

   **b.** Are the samples independent or dependent?

   **c.** If independent, would using dependent samples have been better? Why or Why not? If dependent, explain why dependent samples were used.

**11.** Are females more likely than males to stay in an unsatisfactory job? To answer this question, Foxworthy News randomly sampled 150 females and 170 males. Sixty females said they have stayed in an unsatisfactory job and sixty-five males said they have stayed in an unsatisfactory job. At the 2.5% significance level, what conclusion can be reached?

**12.** Do wearing custom-made shoes, made specifically for the runner, help a runner run faster? To answer this question, twelve competitive runners were randomly selected and their maximum velocity (in mph) was recorded while wearing the custom-made shoes and then again while not wearing the custom-made shoe. (A time for rest was granted and it was randomly determined if they ran with the custom-made shoes first or second. The runners ran 100 yards.)

| Custom shoes | 23.2 | 20 | 17.6 | 22.9 | 21.5 | 23 | 20.8 | 18.9 | 24 | 18.8 | 19.2 | 19.3 |
|---|---|---|---|---|---|---|---|---|---|---|---|---|
| Regular shoes | 14.8 | 16.1 | 15.4 | 18 | 20.4 | 18.2 | 19.3 | 16.3 | 19 | 19.6 | 18.4 | 18.5 |

   **a.** At the 5% significance level, what conclusion can be reached?

   **b.** Are the samples independent or dependent?

   **c.** If independent, would using dependent samples have been better? Why or Why not? If dependent, explain why dependent samples were used.

# CATEGORICAL ANALYSIS

## Section 1: The Chi-Square Goodness of Fit Test:

There are several situations we have already discussed in which the data we are analyzing is count data, such as when testing one or two population proportions. An area of statistics that deals with count data is called *Categorical Analysis*. There are several tests that deal with count data; one such test is the *Chi-Square Goodness of Fit test*.

There are two uses for the Chi-Square Goodness of Fit Test:

1. When testing a distributional assumption about count data. For example, if you have some data and you want to determine if the data came from a population that has a Binomial or a Poisson distribution.

2. When testing more than two population proportions.

We will consider the second use of the Chi-Square Goodness of Fit Test.

## Section 2: Testing More Than Two Population Proportions (The Goodness of Fit Test):

Previously, when we considered other hypothesis testing scenarios, we began with defining the parameter being tested and the point estimate used for testing the parameter. The Chi-Square Goodness of Fit test is not testing an actual parameter; it takes into account what we call the observed and the expected data values. To work the problem, we will follow the same logic as always of hypothesis testing. Thus, in order to know how to work a Goodness of Fit test, we just need to define the observed and expected values, the null and alternative hypothesis, the assumptions, and the test statistic and its distribution.

- The *observed values* are the values actually observed, i.e., our data. We summarize our data (the observed values) into a *one-way table*. If we are testing $k$ population proportions, then there are $k$ cells in the one-way table.

- The *expected values*, denoted as $E_i$, is what we would expect to see if the null hypothesis is true. Thus, $E_i = $ _____. where $i = 1, ..., k$

- The *null and alternative hypothesis*: we can either assume all the proportions are equal or we can assume that the proportions are equal to some value.

For example:

$$H_o: p_1 = p_2 = p_3 = \underline{\hspace{3cm}}$$

Ha: at least one is different than assumed

or

$$H_o: p_1 = .25, p_2 = .55, p_3 = .20$$

Ha: at least one is different than assumed

- The *test statistic*: The test statistic is the *Chi-Square ($\chi^2$) Goodness of Fit Test*, which has the following formula:

$$\chi^2 = \sum \frac{(O_i - E_i)^2}{E_i} \sim \chi^2_{(k-1)}$$

- The *Chi-Square distribution*:

    **1.** There are infinitely many different $\chi^2$ distributions; each one is determined by its degrees of freedom. The degrees of freedom vary, depending on the type of problem being tested.

    **2.** The shape of the distribution depends on the degrees of freedom.

## Chi-Square Distributions

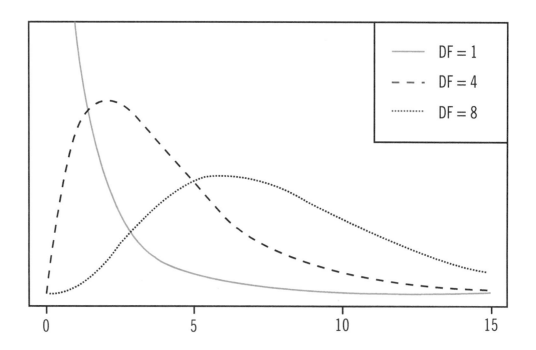

- The assumptions:     1. Have a multinomial experiment

    2. $E_i \geq 5$, for each cell

- The *type of test*: Throughout the test, we always assume that the null hypothesis is true. If Ho is true, what does that imply about $O_i$ and $E_i$?

  Answer:

  What does that imply about the $\chi^2$ value?

  Answer:

  If Ho is false, what does that imply about $O_i$ and $E_i$?

  Answer:

  What does that imply about the $\chi^2$ value?

  Answer:

  Therefore, is the $\chi^2$ Goodness of Fit test a right-tailed test, a left-tailed test, or a two-tailed test?

  Answer:

  Hence, the rejection region will have the form:

  Answer:

And the probability statement for the p-value will always be of the form:

Answer:

- The *Confidence Interval*: The Chi-Square Goodness of Fit test is an overall test. Thus if we reject Ho, that does not imply that all the proportions are different than assumed. It simply implies that at least one proportion is different than assumed.

  If we reject Ho, then we must use *Bonferroni Confidence Intervals* to determine which proportion(s) differ from assumed. To construct a Bonferroni CI, we first define a new alpha,

  $$\alpha^* = \alpha/k.$$

  We then construct a $(1 - \alpha')$% CI for *each* proportion. We use the a $(1 - \alpha^*)$% CI to determine if the proportion is different than assumed. (If there are $k$ proportions being tested, then we will construct $k$ CIs. In reality, we will let JMP do the work for us.)

  For us to conclude using a CI that the proportion is different than assumed, then the assumed value of the proportion must be or must not be contained within the CI?

  Also, although we are having JMP construct $(1 - \alpha^*)$% CIs, we will interpret them as though they are $(1 - \alpha)$% CIs.

- Example 1: Suppose three college students, Sam, Ann, and Tom, are running for Student Body President. Prior to the election, a survey is conducted to determine the voting preferences of a random sample of 300 college students. At the 5% significance level determine if there is a difference in the voter's preference for the three candidates.

|  | *Candidates* |  |
|---|---|---|
| Sam | Ann | Tom |
| 131 | 94 | 75 |

JMP Instructions:

1. One column of the JMP data sheet should contain the counts (frequencies)

   One column of the JMP data sheet should contain the categories, this column must be made a "character"...have to change modeling type to nominal.

2. Under *Analyze*, choose *Distribution*.

3. Click on the column containing the categories and click on *Y*

4. Click on the column containing the frequencies and click on *Freq*.

5. Click *OK*.

6. Under the red options button, select *Test Probabilities*.

7. A "Test Probabilities" section will be added to the output window.

8. Enter the null probabilities in the cells beneath the heading "Hypoth Prob." Note the categories are in alphabetical order.

9. Click *Done*.

10. Use Pearson statistic and p-value.

11. If necessary, choose the CI, make sure and correctly choose your *Bonferroni* level of confidence.

JMP Output:

Distributions

candidate

Frequencies

| Level | Count | Prob |
|-------|-------|---------|
| Ann | 94 | 0.31333 |
| Sam | 131 | 0.43667 |
| Tom | 75 | 0.25000 |
| Total | 300 | 1.00000 |

N Missing

0

3 Levels

Test Probabilities

| Level | Estim Prob | Hypoth Prob |
|-------|------------|-------------|
| Ann | 0.31333 | 0.33333 |
| Sam | 0.43667 | 0.33333 |
| Tom | 0.25000 | 0.33333 |

| Test | ChiSquare | DF | Prob>Chisq |
|------|-----------|-----|-----------|
| Likelihood Ratio | 15.9622 | 2 | 0.0003* |
| Pearson | 16.2200 | 2 | 0.0003* |

Method:

Fix hypothesized values, rescale omitted

Note: Hypothesized probabilities did not sum to 1. Probabilities have been rescaled.

Confidence Intervals

| Level | Count | Prob | Lower CI | Upper CI | 1-Alpha |
|-------|-------|------|----------|----------|---------|
| Ann | 94 | 0.31333 | 0.253245 | 0.380415 | 0.983 |
| Sam | 131 | 0.43667 | 0.369957 | 0.505749 | 0.983 |
| Tom | 75 | 0.25000 | 0.195231 | 0.314136 | 0.983 |
| Total | 300 | | | | |

Note: Computed using score confidence intervals.

# Check Your Understanding

Is the student grade classification at Baylor the same as it is for other universities? In the United States, of those attending a university, the percentage of freshmen students is 25%, the percentage of sophomore students is 16%, the percentage of college juniors is 18%, the percentage of seniors is 23%, and the percentage of graduate or "other" students is 18%. For the 2012/2013 school year, Baylor had 3739 freshmen, 2743 sophomores, 2849 juniors, 3399 seniors, and 2634 graduate or "other" students. At the 1% significance level, can you conclude that the proportions for Baylor differ for the US as a whole?

# ➡ Section 3: Test of Independence:

One of the most frequent uses of the Chi-Square test is for testing whether two criteria of classifications, when applied to a population, are independent.

For example, a market research firm might wish to know whether it can conclude that, for adults in a certain city, the type of beauty care product purchased is associated with the area of residence. This information could help them decide which products to offer and advertise at which store throughout the city.

As before, to work the problem, we just need to know how to arrange the data, the null and alternative hypothesis, the assumptions, and the test statistic and its distribution.

- The *data*: The observed values, i.e., the *data*, is arranged in what is called a *contingency table* or a *two-way table*. The levels of one criterion provide row headings and the levels of the second criterion provide the column headings. The intersection of the rows and columns are called *cells*.

| Characteristic 2 | Characteristic 1 | | |
|---|---|---|---|
| | D | E | F |
| A | 25 | 65 | 42 |
| B | 35 | 22 | 45 |
| C | 62 | 38 | 34 |

- The *hypothesis*:  Ho: independent

  Ha: dependent

- The *test statistic*: The test statistic is the *Chi-Square Test of Independence*, which has the following formula:

$$\chi^2 = \sum \frac{(O_i - E_i)^2}{E_i} \overset{\sim}{} \chi^2_{((r-1)(c-1))}$$

where $r$ = # of rows and $c$ = # of columns containing data.

- Finding $E_i$: Let $P_{AD}$ represent the proportion of people having characteristic A and characteristic D. If the two characteristics are independent, then we know by the multiplication rule for independent events that

$$P_{AD} = P_A \bullet P_D.$$

Thus, since we assume the null hypothesis is true, we know that

$$E_{AD} = P_A \bullet P_D \bullet n$$

Hence, we have the following formula for *Ei*, for each cell:

$$E_i = \frac{(row\ total)(column\ total)}{grand\ total}$$

- The *assumptions*: $E_i \geq 5$, for each cell

- The *type of test*: Throughout the test, we always assume that the null hypothesis is true. By observing the formula for the test statistic, is the $\chi^2$ *Test of Independence* also a right-tailed test, or is it now a left-tailed test, or a two-tailed test? Why?

  Answer:

- Example 1: A market research firm wishes to know whether they can conclude that, for adults in a large metropolitan city, the type of beauty care product purchased is associated with the area of residence. A random sample of 1350 adults from the large metropolitan city was obtained and it was found that of those that live in the city, 143 purchase organic beauty care products, 200 purchase name brand but non-organic beauty care products, and 356 purchase store brand beauty care products. Of the adults that live in the suburbs, 256 purchase organic beauty care products, 175 purchase name brand but non-organic beauty care products, and 220 purchase store brand beauty care products. At the 5% significance level, what conclusion can be reached?

| | Type of Residence | |
|---|---|---|
| Type of Beauty Product | In the City | In the Suburbs |
| Organic | 143 | 256 |
| Name Brand | 200 | 175 |
| Store Brand | 356 | 220 |

- JMP Instructions:

  **1.** Input the levels of one of the characteristics, call it *x*. (Make the column a *character* column.)

  **2.** Input the levels of the other characteristic, call it *y*. (Make the column a *character* column.)

  **3.** Input the counts in the third column. Call it *frequency*.

  **4.** Choose *fit y by x*.

  **5.** Select your *x* and *y* and select *freq* for the data.

  **6.** Hit okay.

  **7.** Under red options triangle by the contingency table, choose *expected*.

- JMP Output:

  ## Contingency Analysis of Area of Residence by Beauty Product

  Freq: frequency

  ## Contingency Table

  Beauty Product by Area of Residence

  | Count Expected | City | Suburb | Total |
  |---|---|---|---|
  | **Name Brand** | 200<br>194.167 | 175<br>180.833 | 375 |
  | **Organic** | 143<br>206.593 | 256<br>192.407 | 399 |
  | **Store Brand** | 356<br>298.24 | 220<br>277.76 | 576 |
  | **Total** | 699 | 651 | 1350 |

  Tests

  | N | DF | -LogLike | RSquare (U) |
  |---|---|---|---|
  | 1350 | 2 | 32.410930 | 0.0347 |

  | Tests | ChiSquare | Prob>ChiSq |
  |---|---|---|
  | Likelihood Ratio | 64.822 | <.0001* |
  | Pearson | 64.155 | <.0001* |

- Comments about the Chi-Square Test of Independence:

   1. For the Chi-Square Test of Independence, the value of the $\chi^2$ test statistic depends on the way the categories are defined for the two variables. If the area of residence is classified as in the city vs. suburb, then the conclusion of the test might be different than if the categories are classified as north side of city, south side of city, east side of city, and west side of city. A report of the test result should identify the categories for the variables.

   2. The Chi-Square Test of Independence treats the classifications as nominal. If the classifications have an understood order (i.e., are ordinal data), the Chi-Square Test of Independence does not take the ordering into account. For example if one characteristic is the "severity" of injury, there is an understood order in the data and one should use a more sophisticated method that takes into account the ordering of the data.

   3. The Chi-Square Test of Independence should not be calculated using percentages or rates; it should only be calculated using observed frequencies.

   4. The Chi-Square Test of Independence is a large sample test and should not be used for small sample sets. If the sample size assumption is not met, you should use *Fisher's Exact Test*.

   5. The Chi-Square test assumes the "levels of the characteristics" are independent. If the columns are dependent samples, such as when each row has the same subjects, then *McNemar's Test* is more appropriate to use.

# Check Your Understanding

A brokerage firm wants to see whether the type of account a customer has between the choices bronze, silver, gold, platinum, or diamond is related to who makes the trade, the customer him/herself or a professional trader hired by the person. A random sample of trades made for its customers over the past year was randomly chosen and it was found that for the trades made by a hired professional broker 55 were at the bronze level, 62 were silver, 75 were gold, 72 were platinum, and 83 were diamond. Of the trades made by the customer him/herself 75 were at the bronze level, 80 were silver, 82 were gold, 65 were platinum, and 67 were diamond. At the 2.5% significance level, what conclusion can be reached?

# ➡ Section 4: Tests of Homogeneity:

In the tests of homogeneity we are testing if several populations are homogeneous or similar with respect to one characteristic.

For example, the local cable provider might want to know whether different age groups are similar (or different) with respect to the type of television programs they prefer to watch.

Once again, to work the problem, we just need to know how to arrange the data, the null and alternative hypothesis, the assumptions, and the test statistic and its distribution.

- The *data*: The observed values, i.e., the *data*, are again arranged in what is called a *contingency table* or a *two-way table*. The levels of the population provide row headings and the levels of the characteristic provide the column headings. The intersection of the rows and columns are called cells.

- The *hypothesis*:    Ho: homogeneous

    Ha: heterogeneous

- The *test statistic*: The test statistic is again the Chi-Square test, which has the same formula:

$$\chi^2 = \sum \frac{(O_i - E_i)^2}{E_i} \sim \chi^2_{((r-1)(c-1))}$$

- Finding $E_i$: Let $E_{AI}$ represent the number of people in age group $I$ that prefer program $A$.

Now $P_A$ represents the proportion of people who prefer program $A$. If all populations are homogneous, then we expect this proportion to be true for all three populations. Thus, since there are $n_i$ people in ethnic group $I$, we have the following formula for $E_{AI}$:

$$E_{AI} = n_i \bullet P_A$$

This is the same as before:    $E_i = \dfrac{(row\ total)(column\ total)}{grand\ total}$

- The *assumptions*: $E_i \geq 5$, for each cell

- The *type of test*: Throughout the test, we always assume that the null hypothesis is true. By observing the formula for the test statistic, is the $\chi^2$ test still a right-tailed test, or is it now a left-tailed test, or a two-tailed test? Why?

Answer:

- Example 1: The local cable provider would like to determine whether different age groups are similar with respect to the type of television programs they prefer to watch. A random sample of adults was obtained. Of those that were less than thirty, 200 preferred reality TV, 185 preferred comedies, 150 preferred dramas, and eighty-eight preferred documentaries. Of those that were between thirty and fifty years of age, ninety-two preferred reality TV, 125 preferred comedies, 188 preferred dramas, and 139 preferred documentaries. Of those that were older than fifty, 52 preferred reality TV, one hundred preferred comedies, 220 preferred dramas, and 150 preferred documentaries. At the 2.5% significance level, what conclusion can be reached?

| Age Group | Type of Show | | | |
| | Reality TV | Comedy | Drama | Documentary |
| --- | --- | --- | --- | --- |
| < 30 | 200 | 185 | 150 | 88 |
| 30 – 50 | 92 | 125 | 188 | 139 |
| > 50 | 52 | 100 | 220 | 150 |

- JMP Instructions:

    1. Input the levels of one of the characteristics, call it *x*. (Make the column a *character* column.)

    2. Input the levels of the other characteristic, call it *y*. (Make the column a *character* column.)

    3. Input the counts in the third column. Call it *frequency*.

    4. Choose *fit y by x*.

    5. Select your *x* and *y* and select *freq* for the data.

    6. Hit okay.

    7. Under red options triangle by the contingency table, choose *expected*.

- JMP Output:

    **Contingency Analysis of Type of Program by Age**

    Freq: frequency 2

    **Contingency Table**

    Age by Type of Program

| Count<br>Expected | Comedy | Documentary | Drama | Reality TV | Total |
|---|---|---|---|---|---|
| < 30 | 185<br>151.231 | 88<br>139.059 | 150<br>205.822 | 200<br>126.887 | 623 |
| > 50 | 100<br>126.714 | 150<br>116.515 | 220<br>172.455 | 52<br>106.316 | 522 |
| 30 − 50 | 125<br>132.054 | 139<br>121.426 | 188<br>179.723 | 92<br>110.797 | 544 |
| Total | 410 | 377 | 558 | 344 | 1689 |

    Tests

| N | DF | -LogLike | RSquare (U) |
|---|---|---|---|
| 1689 | 6 | 74.450206 | 0.0322 |

| Tests | ChiSquare | Prob>ChiSq |
|---|---|---|
| Likelihood Ratio | 148.900 | <.0001* |
| Pearson | 146.160 | <.0001* |

# Check Your Understanding

Campus Life administrators want to determine if students are similar with respect to if and where they work. A random sample of freshman found that 320 do not work, 260 work as a student worker for the University, and 450 work off-campus. A random sample of sophomores found that 280 do not work, 220 work as a student worker for the University, and 350 work off-campus. A random sample of juniors found that 250 do not work, 315 work as a student worker for the University, and 380 work off-campus. A random sample of seniors found that 280 do not work, 220 work as a student worker for the University, and 350 work off-campus. At the 5% significance level what conclusion can be reached?

## ➡ Section 5: The Chi-Square Test Overview:

Fill in the following for testing more than two population proportions:

Parameter:

$E_i$:

General form for Ho:

        Ha:

Test Statistic and its distribution:

Assumptions:

General form of the RR:

General form of probability statement for the p-value:

Method used to determine which proportions differ:

Construct _____  _____  _____ for each $p_i$:

Define $\alpha^\star =$

If the assumed value of $p_i$ is not contained within the CI, then we _____.

Fill in the following for testing independence:

Want to Show:

$E_i$:

General form for Ho:

Ha:

Test Statistic and its distribution:

Assumptions:

General form of the RR:

General form of probability statement for the p-value:

Conclusion if reject Ho:

Conclusion if fail to reject Ho:

Fill in the following for testing homogeneity:

Want to Show:

$E_i$:

General form for Ho:

Ha:

Test Statistic and its distribution:

Assumptions:

General form of the RR:

General form of probability statement for the p-value:

Conclusion if reject Ho:

Conclusion if fail to reject Ho:

# ➡ Chapter 8 Homework:

## Section 2

1. A researcher is interested in determining if the proportion of children that collect specific type of toys is equal. To test, he takes a random sample of 500 children and finds that 36% of them collect Legos, 19% of them collect Beenie Babies, 18% of them collect Webkinz, and 27% of them collect Hot Wheels. At the 10% significance level, what conclusion can be reached?

2. How do the graduation rates for Baylor compare to other US universities? Nationally, 40% of incoming freshman will receive their undergraduate degree within four years, while 13% of incoming freshman will take five years to graduate and 4% will take six years to graduate. Sadly, 43% of all incoming freshman will not graduate within six years time. From a random sample of 20,000 Baylor students who enrolled as freshmen, 10,800 received their undergraduate degree within four years, 3,200 of them took five years to receive their undergraduate degree, and 400 of them took six years to receive their undergraduate degree. Of the 20,000 surveyed, 5,600 of them did not graduate within six years. At the 5% significance level, what conclusion can be reached?

## Section 3

3. Is there a relationship between buying organic produce and having a college degree for males and females? A random sample of 1,077 U.S. adults was obtained. Of those females sampled without a college degree 114 buy organic produce while 212 do not buy organic produce. Of those females sampled with a college degree 125 buy organic produce while 154 do not buy organic produce. Of those males sampled without a college degree, eighty-five buy organic produce while seventy do not buy organic produce. Of those males sampled with a college degree, 165 buy organic produce while 152 do not buy organic produce. At the 2.5% significance level, what conclusion can be reached?

4. For migraine sufferers, is there an association between gender and relief from a migraine with a cold or hot compress? From a random sample of 600 migraine sufferers, 192 females received relief using a hot compress while 208 females received relief using a cold compress. Whereas ninety-eight males received relief using a hot compress and 102 males received relief using a cold compress. At the 1% significance level, what conclusion can be reached?

# Section 4

5. Do males and females differ with respect to the type of degree earned? A random sample of 1,005 undergraduates was obtained. Of the males sampled, 200 were earning a Bachelor of Arts (BA) degree while 280 were obtaining a Bachelor of Science (BS) degree. Of the females sampled, 275 were earning a BA degree while 250 were earning a BS degree. At the 10% significance level, what conclusion can be reached?

6. Are ethnic groups similar with respect to the high school sport their daughters prefer to play? A random sample of US adults whose daughters play a sport in high school was obtained. The data below represents the type of sport each daughter plays and their ethnicity. At the 5% significance level, what conclusion can be reached?

| | Ethnicity | | | |
|---|---|---|---|---|
| Sport | African American | Asian | Caucasian | Hispanic |
| Basketball | 175 | 28 | 120 | 85 |
| Soccer | 55 | 43 | 75 | 165 |
| Softball | 62 | 48 | 90 | 88 |
| Volleyball | 78 | 35 | 88 | 65 |

# Combining the Sections

7. Is there a relationship between gender and the type of car driven? A random sample of 1000 adults was obtained. The subjects sampled were asked their gender and what type of car they drive given choices: car, truck, or SUV. Of the females sampled, 210 drive a car, 110 drive a truck, and 180 drive a SUV. Of the males sampled, 175 drive a car, 140 drive a truck, and 185 drive a SUV. At the 5% significance level what conclusion can be reached? Would your decision change at the 2.5% significance level?

8. Has the percentage of registered nurses earning more advanced academic degrees changed? In 1980 55% of registered nurses had a high school diploma as their highest level of education, 18% had an associate's degree, 22% had a bachelor's degree, and 5% had a master's degree. In 2012, a random sample of 800 registered nurses was obtained. Of those sampled, 111 had a high school diploma as their highest educational degree, 295 had an associate's degree, 289 had a bachelor's degree, and 105 had a master's degree. At the 5% significance level, what conclusion can be reached?

9. Do newly married couples differ with respect to where they want to live? A random sample of newlyweds was obtained. Of the 500 husbands sampled, 135 wanted to live in the suburbs, 298 wanted to live in the city, and sixty-seven wanted to live in the country. Of the 500 wives sampled, 130 wanted to live in the suburbs, 280 wanted to live in the city, and ninety wanted to live in the country. At the 10% significance level, what conclusion can be reached?

# TESTING MORE THAN TWO POPULATION MEANS: ANOVA

When performing a hypothesis test for more than two population means, the *F*-test maximizes power and controls for alpha. The *F*-test, which is an Analysis of Variance (ANOVA) test, is an overall test that determines whether the population means are different by analyzing the variability in the data. The different sources of variability in the data depend upon the structure of the data.

There are three basic principles in determining the design structure. They are as follows:

1.  Random assignment – the subjects/units must be randomly assigned to the treatment/control groups.
2.  Blocking – deals with blocking the subjects/units based on an extraneous factor in order to control for it.
3.  Factorial crossing – deals with comparing a number of treatments including interactions.

There are four basic ANOVA designs. The design used to analyze the data is determined by the design structure. i.e., deciding how to assign the subjects/units to the treatment groups and whether or not the subjects/units will be blocked. The four basic designs are as follows:

1.  The Completely Randomized Design
2.  The Randomized Block Design
3.  The Latin Square Design
4.  The Split Plot/Repeated Measures Design

## ▶ Section 1: ANOVA for the Completely Randomized Design: CR-*k*

In this section we will discuss the design that assumes only one source of variation or factor is analyzed, which is called One-Way ANOVA. We will also assume that the subjects are randomly assigned to the treatment groups. Furthermore, any inference we make applies only to the *k*-treatments. We will not try to draw conclusions to a larger collection of treatments. This is called a *fixed-effects model*.

The Completely Randomized Design is basically an extension of the pooled *t*-test for independent samples.

Each ANOVA design has a model, which is a symbolic representation of the data. A difference in the population means is detected by analyzing the variability in the data. This variability is due to different sources: the variability due to the treatment and the variability just naturally inherent in the data. The different sources of variability are represented in the model.

- Arranging the data:

    $x_{ij}$ ~ the $i^{th}$ observation in the $j^{th}$ treatment

    $\bar{x}_{.i}$ ~ the mean of the $i^{th}$ treatment

    $\bar{x}_{..}$ ~ the mean of all the treatments, called the grand mean.

- The Model – a symbolic representation of a typical value of a data set. For the CR – $k$ design the model is as follows:

$$x_{ij} = \mu + \tau_j + e_{ij} \quad i = 1, ..., n \text{ and } j = 1, ..., k$$

    $\mu$ = the mean of all the $k$ populations

    $\tau_j$ = the difference between the mean of the $j^{th}$ population and the grand mean, called the treatment effect

    $e_{ij}$ = the amount by which an individual measurement differs from the mean of the population to which it belongs, called the error term

Recall that for a Completely Randomized Design we assume we have at least three populations and we are attempting to determine if there is a difference in the means of the populations. In order to see if the treatment is indeed effective, we have to assume that all the groups are as similar as possible. This is what the "$\mu$" represents in the model. We have to assume that the only difference is the treatment. This is what the "$\tau_j$" represents in the model. There is always going to be natural variability in the data. We are assuming that this variability is not due to a bias, but just natural occurring variability in the data. This is what the "$e_{ij}$" represents in the model.

- Assumptions of the model:

    **1.** The data comes from $k$ independent samples

    **2.** The $k$ populations are normally distributed with mean $\mu_j$ & standard deviation $\sigma_j$

    **3.** The population variances are all equal

    **4.** $\tau_j$ are unknown, but $\sum \tau_j = 0$

    **5.** $e_{ij}$ are normally and identically distributed with mean 0 and constant variance, denoted by $\sigma^2$

- Hypothesis for ANOVA: The hypothesis statement can be written in terms of the population means or in terms of the treatment effects. If there is not a treatment effect, then the population means will be equal. Thus the null and alternative hypothesis can be written either way.

    Ho: $\mu_1 = \mu_2 = ... = \mu_k$

    Ha: not all $\mu_{ij}$ are equal

or

    Ho: $\tau_j = 0$ *for all j*

    Ha: $\tau_j \neq 0$ *for some j*

- Calculation of the test statistic: Again, in ANOVA the total variation in the data is portioned into components that are attributable to different sources. (Variation refers to the sum of squared deviations from the mean or for short – Sum of Squares – *SS*.)

For a *CR-k* design, the total variation is portioned into two components: the treatment and the error. This implies SST = SSW + SSB where

$$SST = \sum_{j=1}^{k} \sum_{i=1}^{n_j} (x_{ij} - \bar{x}_{..})^2$$

$$SSW = \sum_{j=1}^{k} \sum_{i=1}^{n_j} (x_{ij} - \bar{x}_{.j})^2$$

$$SSB = \sum_{j=1}^{k} n_j (\bar{x}_{.j} - \bar{x}_{..})^2$$

Picture 1: Represents the different Sums of Squares when Ho is true.

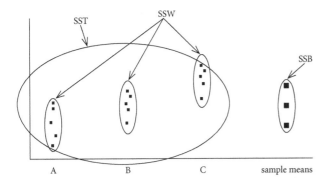

Picture 2: Represents a possible example of the different Sums of Squares when Ho is false.

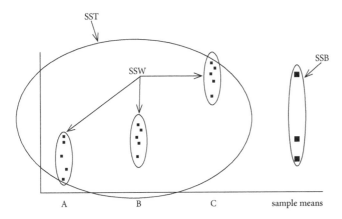

Recall that we are testing if the means are equal. If the sample means are clustered together, then this would tend to support Ho. On the other hand, a great deal of variability between the sample means would suggest support for Ha.

In the ANOVA method, when saying that the variability among the sample means is large, we are saying that the variability *between* the sample means (due to the "treatment") is large in relation to the variability *within* the samples. ** Remember, we are assuming the variability within each population is the same.

Thus ANOVA considers two estimates of $\sigma^2$:

1. MSW which measures the variability *within* the samples:

$$MSW = \frac{SSW}{df} = \frac{\sum\limits_{j=1}^{k}\sum\limits_{i=1}^{n_j}(x_{ij} - \bar{x}_{\cdot j})^2}{N-k}$$

2. MSB which measures the variability *between* the samples:

$$MSB = \frac{SSB}{df} = \frac{\sum\limits_{j=1}^{k}n_j(\bar{x}_{\cdot j} - \bar{x}_{\cdot\cdot})^2}{k-1}$$

To analyze the variability in the data, ANOVA compares the above two estimates for $\sigma^2$. This works because of the following truths about the above estimates:

- *MSW* gives a valid estimate for $\sigma^2$ whether Ho is true or not.
- *MSB* is a valid estimate for $\sigma^2$ only if Ho is true.
- It can be shown that the MSB estimate is an inflated estimate for $\sigma^2$ if Ho is false.

Thus, to make our decision, we compare the two estimates for $\sigma^2$ by considering the ratio of the two estimates. This leads us to the test statistic:

$$F = \frac{MSB}{MSW} \sim F\text{-distribution with } (k - 1, N - k) \text{ degrees of freedom.}$$

If Ho is true, what does that imply about the value of $F$?

Answer:

If Ho is false, what does that imply about the value of $F$?

Answer:

Thus, the *F*-test is a _____ tailed test.

Hence, the probability statement for the p-value will have the form:

Answer:

- Properties of the *F*-dist.:

  1. Each *F*-distribution has a pair of degrees of freedom:

     $k - 1$ = degrees of freedom for the numerator and $N - k$ = degrees of freedom for the denominator

  2. The graph of the *F*-distribution starts at zero and extends indefinitely to the right. It is a skewed-right distribution.

- The information for ANOVA is typically organized in an ANOVA table:

| Source | Degrees of Freedom | Sums of Squares | Mean Squares | F-ratio |
|--------|-------------------|-----------------|--------------|---------|
| Between Samples | $k - 1$ | SSB | $MSB = \dfrac{SSB}{k-1}$ | $\dfrac{MSB}{MSW}$ |
| Within Samples | $N - k$ | SSW | $MSW = \dfrac{SSW}{N-k}$ | |
| Total | $N - 1$ | SST | | |

- Tukey's Test:

  Since the *F*-test is an overall test, if we reject Ho, we do not know which means are different. One possible method for determining which specific means differ is Tukey's HSD (Honestly Significant Difference) test, which is a multiple comparison test.

  Tukey's test compares each pair of means to a calculated single value called the HSD value (or LSD value depending on whether the sample sizes are equal or not). This value represents the minimum distance the sample means need to be apart in order to infer that the population means are indeed different. Thus, the HSD value is compared to the absolute value of the difference between the means. For example, if comparing $\mu_1$ versus $\mu_2$ then the following difference is compared:

$$|\bar{x}_2 - \bar{x}_1| - HSD$$

  If the absolute difference of the means is larger than the HSD value (i.e., the above expression is positive), then the means are said to be statistically different.

The HSD value is found by the formula: HSD $= q_{\alpha,k,N-k}(MSE/n)^{1/2}$

where

$k = $ # of populations

$N = $ total # of observations

$n = $ # of observations in the sample

The $q$ value is known as the Studentized Range Statistic and is equal to

$$q = \frac{\overline{x}_{max} - \overline{x}_{min}}{\dfrac{MSE}{n}}$$

Tukey's test assumes $n$ is the same for each sample. If $n$ is not the same, then Tukey's test is conservative.

Example 1: Suppose we have three different allergy drugs and we wish to determine if there is a difference between how long the "medicine head" feeling lasts for all adults who are under the influence of the drug. At the 5% significance level do the data suggest that the average amount of time the medicine head feeling lasts is different while under the influence of the drugs?

| A | 18 | 20 | 21 | 23 | 19 | 17 | 16 |
| B | 21 | 26 | 26 | 27 | 25 | 24 | 23 |
| C | 30 | 24 | 26 | 25 | 30 | 29 | 27 |

**Hypothesis:**

Ho:

Ha:

**Observed Value:**

**P-value:**

**Decision:**



**If appropriate, use Tukey's Test to determine which means differ:**







**Conclusion:**

- JMP directions:
  1. Put all the data in one column
  2. Under the *columns* menu, choose *new column*
  3. Under *data type*, choose *character*
  4. Use this column to label the data as 1, 2, 3, etc.
  5. Under *analyze*, choose *fit y by x*
  6. Click on *Column 1*, then click on the *Y, Response* button.
     Click on *Column 2*, then click on *X, Factor* button
  7. Choose *Normal Quantile* plot to check normality
  8. Choose *Unequal Variances* to check equal variances
  9. Choose means *Anova*, to get the ANOVA table
  10. If needed, choose *Compare Means, All-pairs Tukey, HSD*, to do Tukey's test

- JMP output:

**Oneway Analysis**

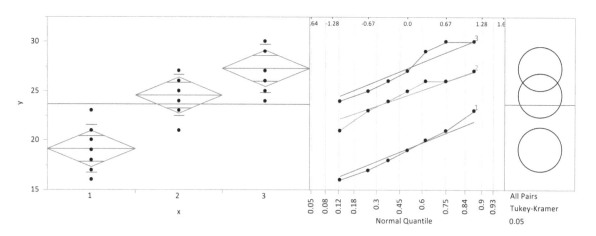

**Oneway Anova**

**Analysis of Variance**

| Source | DF | Sums of Squares | Mean Square | F-ratio | Prob > F |
|--------|----|-----------------|-------------|---------|----------|
| x | 2 | 240.66667 | 120.333 | 22.5625 | <.0001* |
| Error | 18 | 96.00000 | 5.333 | | |
| C. Total | 20 | 336.66667 | | | |

**Tests that the Variances are Equal**

| Level | Count | Std Dev | MeanAbsDif to Mean | MeanAbsDif to Median |
|-------|-------|---------|--------------------|--------------------|
| 1 | 7 | 2.410295 | 1.877551 | 2.000000 |
| 2 | 7 | 2.070197 | 1.632653 | 1.714286 |
| 3 | 7 | 2.429972 | 2.040816 | 2.142857 |

| Test | F Ratio | DFNum | DFDen | Prob > F |
|------|---------|-------|-------|----------|
| O'Brien[.5] | 0.1650 | 2 | 18 | 0.8491 |
| Brown-Forsythe | 0.2958 | 2 | 18 | 0.7475 |
| Levene | 0.2262 | 2 | 18 | 0.7998 |
| Bartlett | 0.0877 | 2 | . | 0.9161 |

**Means Comparisons**

**Comparisons for all pairs using Tukey-Kramer HSD**

**Confidence Quantile**

| $q^*$ | Alpha |
|-------|-------|
| 2.55216 | 0.05 |

**SD Threshold Matrix**

| Abs(Dif)-HSD | 3 | 2 | 1 |
|--------------|-----|-----|-----|
| 3 | -3.1505 | -0.4362 | 4.9924 |
| 2 | -0.4362 | -3.1505 | 2.2781 |
| 1 | 4.9924 | 2.2781 | -3.1505 |

Positive values show pairs of means that are significantly different.

# Check Your Understanding

With the increase in health insurance cost, many healthy Americans are looking into different healthcare plans. Americans can choose from among four different health plans. It is of interest to know whether the mean yearly co-pay cost is different for the four different health plans for healthy Americans. To test, four random samples of healthy Americans were selected, where each sample represents one of the health plans, and the yearly co-pay cost is listed below. At the 5% significance level, what conclusion can be reached?

| Plan A: | 1198 | 1195 | 1194 | 1203 | 1204 | 1205 | 1206 | 1199 | 1207 | 1200 |
|---------|------|------|------|------|------|------|------|------|------|------|
| Plan B: | 1227 | 1222 | 1234 | 1218 | 1225 | 1227 | 1228 | 1226 | 1212 | 1229 |
| Plan C: | 1228 | 1230 | 1239 | 1237 | 1233 | 1234 | 1238 | 1232 | 1229 | 1218 |
| Plan D: | 1216 | 1205 | 1212 | 1214 | 1211 | 1205 | 1199 | 1202 | 1208 | 1203 |

# ➡ Section 2: ANOVA for the Randomized Block Design: RB-*k*

Recall that an extraneous factor (also known as a *nuisance* or *confounding factor*) is anything that affects the outcome of the experiment other than the treatment. In terms of the variability in the data, an extraneous factor is a source of variability due to a bias and not just the variability due to the treatment and the natural variability that occurs in the data. Recall that in a Completely Randomized Design, ANOVA compares the variability due to the treatment versus the natural variability inherent in the data. If there is an extraneous factor, then the non-treatment variability is more than just "natural variability" in the data. Uncontrolled extraneous factors can bias the results or at the least distort the effects of the conditions. Uncontrolled factors can also make the effects impossible to isolate because the bias in the data due to the extraneous factor will increase the variability in the data that is not contributable to the treatment. Thus this variability needs to be modeled or controlled for in the experiment. The Randomized Block Design controls for the variability due to the confounding factor by including that factor into the model.

The Randomized Block Design is an extension of the paired-*t* test.

The Randomized Block Design "blocks" the units based on the confounding factor. This is done by the following steps:

1.  Sort the subject/units into groups (blocks) of similar characteristics of the confounding factor. The number of units in each block must equal the number of treatments

2.  Randomly assign the different treatments to the subjects/units separately within each block. Note that each unit gets one treatment and each block of subject/units gets the complete set of treatments.

It is important to note that the effect of blocking depends on the way the units are sorted into the blocks. It is of upmost importance that the blocks should be as similar as possible in terms of what "matters the most" to the response. i.e., you should be blocking based on the specific confounding factor that most affects the outcome. For example, if testing different sunscreens, you would block based on skin type, not on hair color.

*   Arranging the data:

    $x_{ijl}$ ~ the $i^{th}$ observation in the $j^{th}$ treatment in the $l^{th}$ block

    $\bar{x}_{\cdot j \cdot}$ ~ the mean of the $j^{th}$ treatment

    $\bar{x}_{\cdot \cdot l}$ ~ the mean of the $l^{th}$ block

    $\bar{x}_{\cdot \cdot \cdot}$ ~ the mean of all the data, called the grand mean.

*   The Model:

$$y_{ijl} = \mu + \tau_j + \beta_l + e_{ijl}$$

The terms $\mu$, $\tau_j$, and $e_{ijl}$ are defined as in the Completely Randomized Design. The term $\beta_l$ represents the block effect, which is the variability in the data due to the extraneous factor.

The ANOVA test for the Randomized Block Design will analyze the variability in the data by portioning the total variability into three components: variability due to the treatment (which is represented by the term $\tau_j$ in the model), variability due to the block (which is represented by the term $\beta_l$ in the model), and the natural variability in the data (which is represented by the term $e_{ijl}$).

$$SST = SSTr + SSBl + SSE$$

This information can be organized in the ANOVA table.

ANOVA Table:

| Source | Degrees of Freedom | Sums of Squares | Mean Squares | F-ratio |
|---|---|---|---|---|
| **Treatment** | $k - 1$ | $SSTr$ | $MSTr = \dfrac{SSTr}{k-1}$ | $F = \dfrac{MSTr}{MSE}$ |
| **Block** | $b - 1$ | $SSBl$ | $MSBl = \dfrac{SSBl}{b-1}$ | $F = \dfrac{MSBl}{MSE}$ |
| **Error** | $(k-1)(b-1)$ | $SSE$ | $MSE = \dfrac{SSE}{(k-1)(b-1)}$ | |
| **Total** | $N - 1$ | $SST$ | | |

- Hypothesis: The Randomized Block Design can test the two different sources of variability: the variability due to the treatment (i.e., the treatment effect) and the variability due to the blocks (i.e., the block effect). This gives two hypotheses to test:

The hypothesis for testing the *treatment effect* can be written in terms $\mu$ or $\tau$. This hypothesis tests the equality of the means.

$Ho: \mu_1 = \mu_2 = \ldots = \mu_k$      or      $Ho: \tau_j = 0$ *for all j*

$Ha:$ *not all the means are equal*          $Ha: \tau_j \neq 0$ *for some j*

Just as with the Completely Randomized Design, concluding there is no treatment effect is equivalent to concluding that the population means are equal. Furthermore, as with the Completely Randomized Design, if this hypothesis is rejected, Tukey's test is used to determine which means differ.

The hypothesis for testing the *block effect* is testing whether there is indeed a block effect. This can be thought of as testing whether you gained anything by including the confounding factor in the model.

$Ho: \beta_l = 0$ *for all l*

$Ha: \beta_l \neq 0$ *for some l*

If this hypothesis is not rejected, then nothing was gained by performing a Randomized Block Design. One should have simply performed a Completely Randomized Design.

- Assumptions: The assumptions are the same as for the Completely Randomized Design (except for independent samples) and are verified with the same methods as previously discussed in Section 1.

- Example 1: Researchers are interested in determining if diet and exercise added to medication provide an effective method for controlling high systolic blood pressure. The researchers tested four different treatments: medication alone, medication with controlled diet, medication combined with controlled diet and regular exercise, and diet and exercise alone. It is also believed that an individual's BMI level is a confounding factor for systolic blood pressure, so the researchers blocked based on four different BMI levels: below normal, normal, obese, and morbidly obese. The table below gives the systolic blood pressure for the different treatments and blocks.

| | Medication Alone | Medication with Diet | Medication with Diet and Exercise | Diet and Exercise Alone |
|---|---|---|---|---|
| Below BMI | 123 | 121 | 117 | 122 |
| Normal BMI | 118 | 117 | 116 | 121 |
| Obese BMI | 125 | 123 | 119 | 124 |
| Morbidly Obese BMI | 122 | 118 | 117 | 123 |

At the 5% significance level, is there a difference in the average systolic blood pressure for the four groups? This tests the treatment effect:

**Hypothesis:**

Ho:

Ha:

**Observed Value:**

**P-value:**

**Decision:**

**If appropriate, use Tukey's Test to determine which means differ:**

Conclusion:

**b.** At the 5% significance level, did the researchers gain power by blocking the subjects based on BMI level? This tests the block effect:

**Hypothesis:**

Ho:

Ha:

**Observed Value:**

**P-value:**

**Decision:**


**Conclusion:**


JMP Instructions:

1. Enter the data as shown below. Make sure the columns for the labels (for example, for BMI level and treatment level) are made character.

| BMI | Treatment | Pressure |
|-----|-----------|----------|
| 1 | 1 | 123 |
| 1 | 2 | 121 |
| 1 | 3 | 117 |
| 1 | 4 | 122 |
| 2 | 1 | 118 |
| 2 | 2 | 117 |
| 2 | 3 | 116 |
| 2 | 4 | 121 |
| 3 | 1 | 125 |
| 3 | 2 | 123 |
| 3 | 3 | 119 |
| 3 | 4 | 124 |
| 4 | 1 | 122 |
| 4 | 2 | 118 |
| 4 | 3 | 117 |
| 4 | 4 | 123 |

2. Under *Analyze*, choose *fit y by x*.

3. Put the pressure as *y*, the treatment as *x*, and the BMI as *block*.

4. Hit *ok*.

5. Use the same options to verify assumptions and obtain the output as you do for a CR-$k$ design.If needed, choose Compare Means, *All-pairs Tukey, HSD*, to do Tukey's test

JMP Output:

## Oneway Analysis of Pressure by Treatment

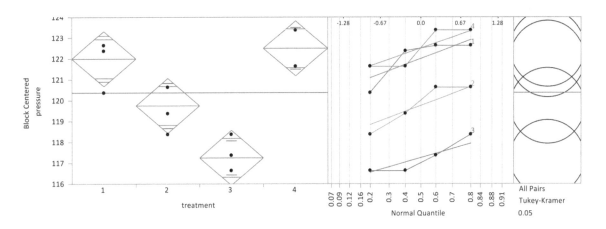

## Oneway Anova

### Analysis of Variance

| Source | DF | Sums of Squares | Mean Square | F-ratio | Prob > F |
|---|---|---|---|---|---|
| treatment | 3 | 69.25000 | 23.0833 | 16.9592 | 0.0005* |
| block | 3 | 46.25000 | 15.4167 | 11.3265 | 0.0021* |
| error | 9 | 12.25000 | 1.3611 | | |
| C. Total | 15 | 127.75000 | | | |

### Tests that the Variances are Equal

| Level | Count | Std Dev | MeanAbsDif to Mean | MeanAbsDif to Median |
|---|---|---|---|---|
| 1 | 4 | 1.089725 | 0.8125000 | 0.6250000 |
| 2 | 4 | 1.089725 | 0.8750000 | 0.8750000 |
| 3 | 4 | 0.829156 | 0.6250000 | 0.6250000 |
| 4 | 4 | 1.010363 | 0.8750000 | 0.8750000 |

| Test | F Ratio | DFNum | DFDen | Prob > F |
|---|---|---|---|---|
| O'Brien[.5] | 0.1463 | 3 | 12 | 0.9301 |
| Brown-Forsythe | 0.2222 | 3 | 12 | 0.8791 |
| Levene | 0.3496 | 3 | 12 | 0.7902 |
| Bartlett | 0.0815 | 3 | . | 0.9701 |

**Means Comparisons**

**Comparisons for All Pairs using Tukey-Kramer HSD**

**Confidence Quantile**

| q* | Alpha |
|---|---|
| 3.12180 | 0.05 |

**LSD Threshold Matrix**

| Abs(Dif)-HSD | 4 | 1 | 2 | 3 |
|---|---|---|---|---|
| 4 | -2.5754 | -2.0754 | 0.1746 | 2.6746 |
| 1 | -2.0754 | -2.5754 | -0.3254 | 2.1746 |
| 2 | 0.1746 | -0.3254 | -2.5754 | -0.0754 |
| 3 | 2.6746 | 2.1746 | -0.0754 | -2.5754 |

Positive values show pairs of means that are significantly different.

# Check Your Understanding

Due to the increase in monthly healthcare premiums, many healthy Americans are investigating to see which health care system has the lowest monthly premiums. The premiums are determined by the number of members in the family. The data below is the monthly premiums for families for four different health plans.

| | Plan A | | Plan B | | Plan C | | Plan D | |
|---|---|---|---|---|---|---|---|---|
| **2 Family Members** | 1150 | 1145 | 1165 | 1154 | 1121 | 1133 | 1159 | 1163 |
| **3–5 Family Members** | 1345 | 1401 | 1420 | 1505 | 1431 | 1401 | 1390 | 1436 |
| **6+ Family Members** | 1522 | 1602 | 1599 | 1654 | 1642 | 1603 | 1684 | 1699 |

**a.** At the 2.5% significance level, is there a difference in the average monthly premiums for the four different health plans.

**b.** Test to determine if there is a benefit to blocking based on family size.

# ➡️ Chapter 9 Homework:

# Section 1

1.  What method works best for removing dirt from laundry? Researchers used three different methods to wash equally soiled laundry: soaking clothes in tepid water, soaking clothes in ice cold water, not soaking clothes at all. The amount of dirt left after washing by each method was measured using sensors and is listed in the table. At the 1% significance level, is there a difference in the average amount of dirt left?

| Soaked with Ice | 0.34 | 0.78 | 0.94 | 1.01 | 0.73 | 1.21 | 0.81 | 0.70 |
|---|---|---|---|---|---|---|---|---|
| Soaked in Water | 1.16 | 1.25 | 1.43 | 1.48 | 1.49 | 1.36 | 1.71 | 1.77 |
| Not Soaked | 1.50 | 1.81 | 1.70 | 1.35 | 1.29 | 1.66 | 1.53 | 1.64 |

2.  Does the type of prenatal care affect a baby's birth weight? Researchers compared the birth weight of babies following four different types of prenatal care: no prenatal care at all, over-the counter-prenatal vitamins, prescribed prenatal vitamins with irregular doctor visits, prescribed prenatal vitamins with regular doctor visits. At the 5% significance level, what conclusions can be reached?"

| No Prenatal Care | 5.9 | 5.8 | 6.2 | 5.6 | 5.7 | 6.1 | 6.0 |
|---|---|---|---|---|---|---|---|
| Over-the-Counter Vitamins Only | 6.0 | 6.1 | 6.1 | 6.3 | 6.5 | 6.3 | 6.2 |
| Irregular Doctor Visits | 7.0 | 6.7 | 6.9 | 7.1 | 7.2 | 7.1 | 6.8 |
| Regular Doctor Visits | 7.1 | 7.6 | 7.5 | 6.9 | 7.4 | 7.2 | 7.3 |

# Section 2

3.  Does an outfit really improve a professional swimmer's time? To test this, ten professional swimmers swam a 100 m race wearing three different outfits: a full-body water repellent suit, water-repellent trunks, and non-water repellent trunks. The table below gives the swimmers' time in seconds. At the 5% significance level, determine if the outfit does indeed make a difference. Also test to determine if the swimmer was a confounding factor.

| Swimmer | 1 | 2 | 3 | 4 | 5 | 6 | 7 | 8 | 9 | 10 |
|---|---|---|---|---|---|---|---|---|---|---|
| Full-Body Water Repellent Suit | 49.80 | 50.52 | 49.36 | 50.02 | 50.36 | 49.89 | 50.31 | 50.05 | 50.81 | 49.73 |
| Water-Repellent Trunks | 51.07 | 53.47 | 52.91 | 53.07 | 53.11 | 52.89 | 53.14 | 53.07 | 53 | 52.68 |
| Non-Water Repellent Trunks | 53.66 | 54.91 | 54.55 | 53.85 | 54.63 | 55.01 | 55.39 | 53.93 | 55.69 | 55.21 |

4. With the increase in online courses, university researchers are interested to learn if there is a difference in the average overall grade for students taking an online statistics course with and without video supplements versus students taking a traditional in-class statistics course. Since it is believed that a student's previous statistical knowledge can affect the outcome, the researchers blocked the students based on their previous statistical knowledge from taking AP statistics: scored 4 or above on the AP Statistics exam, scored below a 3 on the AP Statistics exam, or did not take AP Statistics. The table below gives the student's overall grades.

| | Treatments | | |
| --- | --- | --- | --- |
| *Blocks* | *In-Class Course* | *On-line with Video* | *On-line without Videos* |
| *Scored 4 or Above on the AP Statistics Exam* | 90 93 89 86 88 90 | 82 84 85 88 79 83 | 79 76 78 75 77 81 |
| *Scored 3 or Below on the AP statistics Exam* | 91 89 90 93 95 92 | 86 87 82 82 89 88 | 81 83 86 82 79 76 |
| *Did not Take AP Statistics* | 93 89 97 88 92 86 | 86 88 84 87 89 85 | 78 76 80 77 75 79 |

a. Test whether the average overall scores differ for the three different types of statistic courses.

b. Is there a benefit to blocking on the student's previous statistical knowledge? Justify your answer.

# SIMPLE LINEAR REGRESSION

**CHAPTER 10**

## ➡ Section 1: Introduction:

In regression, we look to see what type of relationship exists between variables and use the relationship to build a model that can be used for describing, controlling, and predicting.

- A *mathematical* relationship versus a *statistical* relationship:

  In a mathematical relationship, we have an exact relationship between variables: if we know the value of one variable, there is a formula that will give us the value of the other variable.

  In a statistical relationship, we have an approximate relationship between variables: if we know the value of one variable, there is a model that will give us an estimate for the value of the other variable. The stronger the relationship, the better estimates the model will give.

- The type of model built depends upon the type of variables and the type of relationship that exists between the variables.

- Consider the following scatterplots:

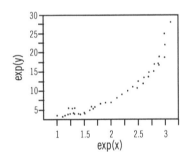 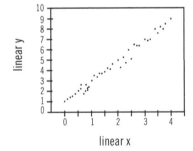 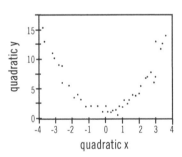

## ➡ Section 2: Basic Properties and Definitions:

- $X$ is called the "independent" or "predictor" variable
- $Y$ is called the "response" or "dependent" variable

- Observed Values: The observed values are the actual known values of $Y$, given $X$.

    Notation:

- Predicted Values: The predicted values, also called the fitted values, are the values of each case that is predicted (or found by) the regression equation.

    Notation:

- Residuals: The residuals are the difference between the observed values and the predicted values.

    Notation:

# ➡ Section 3: Simple Linear Regression:

In Simple Linear Regression, there is one response variable ($Y$), one predictor variable ($X$) and the relationship between the response and predictor variable is linear.

The hypothetical Regression Model: $\qquad Y = \beta_0 + \beta_1 X + \varepsilon$

where

- $\beta_0$ represents the _____ .

- $\beta_1$ represents the _____ .

- $\varepsilon$ represents the _____ .

- $\varepsilon$ is what makes the relationship _____ .

- In regression, for each value of $x$, there is a subpopulation of $y$ values. We use the notation $\mu_{Y|x}$ to denote the average value of all $Y$'s given $X = x$. On average, the random errors are assumed to be zero.

    - Thus, the average value of $Y$ given $X = x$ is: $\qquad \mu_{Y|x} = \beta_0 + \beta_1 x.$

    Hence, the regression line is predicting the *average* value of y given x.

- The graph below depicts what is meant by for each $x$ there is a subpopulation of $y$ values and what we mean by predicting the average value of $y$:

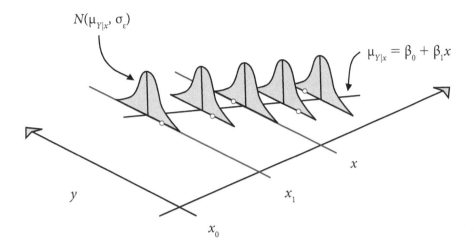

$N(\mu_{Y|x}, \sigma_\varepsilon)$

$\mu_{Y|x} = \beta_0 + \beta_1 x$

$x$

$y$

$x_1$

$x_0$

*From Dr. Seaman's STA 3381 notes.*

- In practice, we sample for each $x$, one $y$. A picture of this is called a scatterplot:

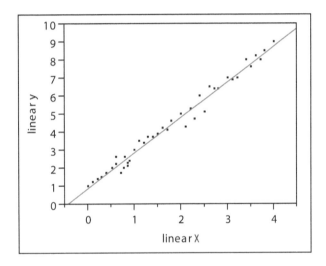

- The sample regression line is the line, $\hat{y} = b_0 + b_1 x$, (also called the line of best fit) where

$b_0$ represents _____ .

$b_1$ represent: _____ .

- What is the relationship between the population regression line and the sample regression line? The parameters in the hypothetical model need to be estimated...

$$\mu_{Y|x} = \beta_0 + \beta_1 x$$

$$\hat{y} = b_0 + b_1 x$$

- $\hat{y}$ is a _____ for $\mu_{Y|x}$

- $b_0$ is a _____ for $\beta_0$

- $b_1$ is a _____ for $\beta_1$

 

- To find the sample regression line we want to choose the line that best "fits" the data. What exactly is meant by "best fits the line"?

  It means, we choose the line that minimizes the residuals. Recall,

  Residual = observed value − predicted value = $y - \hat{y}$

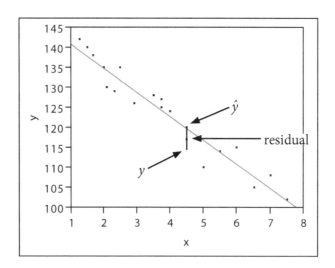

Each point has a residual, so to find the regression line, we want to find the line that minimizes *all* the residuals, which you might think would lead to minimizing the equation:

$$\sum_{i=1}^{n}(y_i - \hat{y}_i)$$

But, what will this sum always equal?

Answer:

 

Why?

Answer:

Thus, we have to minimize the equation:

$$\Sigma_{i=1}^{n}(y_i - \hat{y}_i)^2 = \Sigma_{i=1}^{n}(y_i - (b_0 + b_1 x_i))^2$$

To minimize the equation, we use differential calculus, i.e., we take the derivative with respect to $b_0$ and with respect to $b_1$, set equal to zero and solve. This gives us the following sample regression line.

- $\hat{y} = b_0 + b_1 x$ where $b_0 = \bar{y} - b_1 \bar{x}$ and $b_1 = \dfrac{SS_{xy}}{SS_{xx}}$
  and
  $SS_{xx} = \Sigma_{i=1}^{n} x_i^2 + \dfrac{1}{n}(\Sigma_{i=1}^{n} x_i)^2$ and $SS_{xy} = \Sigma_{i=1}^{n} x_i y_i + \dfrac{1}{n}(\Sigma_{i=1}^{n} x_i)(\Sigma_{i=1}^{n} y_i)$.

- We will use JMP to find the regression line, but we need to know how to we interpret the values.

- *What is the interpretation of the slope?*

  Answer:

- *What is the interpretation of the intercept?* The intercept may not always have an interpretation. When it does have an interpretation, it means:

  Answer:

  When will it not have an interpretation?

  Answer:

  Extrapolation: When the model is used to estimate $y$ for values of $x$ that are "outside" the range of values considered.

- Simple Linear Regression Example:

Is there a relationship between time one spends in strenuous exercise and blood pressure? Researchers would like a model that will estimate systolic blood pressure given amount of time (in hours) a person spends in strenuous exercise per week. To create the model and hence answer the question, a random sample of twenty adults was obtained. Use the data to determine if the relationship is linear and if so, to build the model.

| Time | Pressure | Time | Pressure |
|------|----------|------|----------|
| 1.67 | 138 | 2 | 135 |
| 1.5 | 140 | 2.3 | 129 |
| 2.92 | 126 | 3.5 | 128 |
| 2.08 | 130 | 2.5 | 135 |
| 1.25 | 142 | 3.75 | 127 |
| 4.5 | 120 | 3.75 | 125 |
| 5 | 110 | 6 | 115 |
| 7 | 108 | 4 | 124 |
| 5.5 | 114 | 6.5 | 105 |
| 4.5 | 117 | 7.5 | 102 |

- Which variable is the independent variable?

  Answer:

- Which variable is the dependent variable?

  Answer:

- Is the relationship linear?

  Answer:

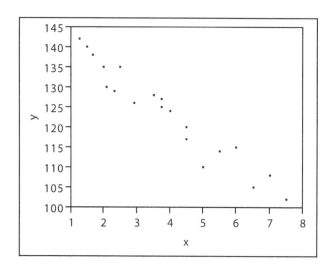

- What is the regression line?

  Answer:

- Interpret the coefficients of the regression line:

  Answer:

- Use the model to estimate the systolic blood pressure for someone who exercises an hour and a half five days a week.

  Answer:

- Interpret the estimate found above:

- Instructions for using JMP to analyze the data:
  1. Enter the data into two separate columns.
  2. Click on the dependent variable and click on the *Y, Response* button. Click on the independent variable and click on the *X, Factor* button. Click *okay.*
  3. Choose *Fit Y by X* under the *Analyze* window.
  4. Using the options button, choose *fit line.*

- JMP Output:

## *Bivariate Fit of y By x*

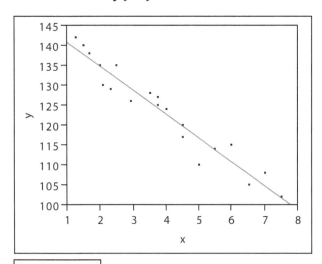

——— Linear Fit

### *Linear Fit*

y = 146.8355 - 6.0050193*x

### *Summary of Fit*

| | |
|---|---|
| RSquare | 0.934533 |
| RSquare Adj | 0.930895 |
| Root Mean Square Error | 3.102798 |
| Mean of Response | 123.5 |
| Observations (or Sum Wgts) | 20 |

### *Parameter Estimates*

| Term | Estimate | Std Error | t Ratio | Prob>ltl |
|---|---|---|---|---|
| Intercept | 146.8355 | 1.612659 | 91.05 | <.0001* |
| x | -6.005019 | 0.374622 | -16.03 | <.0001* |

# Check Your Understanding

A pharmaceutical company has developed a new drug for patients with Wolfe-Parkinson-White syndrome designed to control paroxysmal supraventricular tachycardia (PSVT). Patients with PSVT have a heart rate of more than 100 beats per minute. (BPM) Suppose you are interested in modeling the relationship between the decrease in pulse rate in BPM and dosage in cc. Different doses were administered to 16 randomly selected patients and 20 minutes later the decrease in each patient's pulse rate was recorded. The data collected is given below.

| Dosage | 2 | 4 | 1.5 | 1 | 3 | 3.5 | 2.5 | 3 | 3.5 | 1.5 | 2 | 2 | 3.75 | 3.5 | 1.75 | 1.25 |
|--------|---|---|-----|---|---|-----|-----|---|-----|-----|---|---|------|-----|------|------|
| Drop in HR | 14 | 24 | 11 | 6 | 18 | 22 | 17 | 15 | 22 | 9 | 16 | 14 | 23 | 25 | 12 | 12 |

**a.** Which variable is the independent variable?

**b.** Which variable is the dependent variable?

**c.** Is a linear model appropriate? Explain your answer.

**d.** If a linear model is appropriate, find the line of best fit.

**e.** If appropriate, interpret the coefficients of the model.

**f.** If appropriate, estimate the drop in pulse rate for a dosage level of 2.75 cc. Interpret the estimate.

**g.** If appropriate, estimate the drop in pulse rate for a dosage level of 5.5 cc. Interpret the estimate.

# ➡ Section 4: Confidence Intervals:

For estimating parameters, the regression line, or "line of best fit," is, in essence, a point estimate for the parameters. i.e.,

$$\mu_{Y|x} = \beta_0 + \beta_1 x$$
$$\uparrow \quad \uparrow \quad \uparrow$$
$$\hat{y} = b_0 - b_1 x$$

Instead of just using the point estimates, we might want more of an idea as to what we can expect to see from the data. Using just a point estimate doesn't take into account the variability in the data. Thus we might want to calculate a confidence interval for the slope, $\beta_1$, and the average of $Y$, given $x$, $\mu_{Y|x}$.

- *Confidence Interval for the Population Slope in Simple Regression*: To have more of an idea about the average change in $Y$ for a unit change in $x$, we need a confidence interval for the slope, $\beta_1$.

  Recall, the general form of a CI, is as follows:

$$\text{pt. est.} \pm \text{RF*SE}$$

  Filling in the values, we get:

We use $n - 2$ degrees of freedom for the $t$-table and obtain the standard error of the slope, $S_{b_1}$, from the JMP output.

**Parameter Estimates**

| Term | Estimate | Std Error | t Ratio | Prob>ltl |
|------|----------|-----------|---------|----------|
| Intercept | 146.8355 | 1.612659 | 91.05 | <.0001* |
| x | -6.005019 | 0.374622 | -16.03 | <.0001* |

- Example 1: In the blood pressure vs. exercise time example find a 95% CI for the population slope:
  - pt. est.:
  - SE:

  - RF:

  So a 95% CI for $\beta_1$ is:

- Interpretation:

  We are 95% confident that if we _____ the amount of time spent in strenuous exercise by _____, the _____ systolic blood pressure will _____ between _____ and _____ mmhg, on average.

- Using the CI on $\beta_1$ to verify if there is a linear relationship between the variables:

  Since the CI does not contain the value _____, we can conclude that there is a linear relationship between hours per week of strenuous exercise and systolic blood pressure levels.

- Confidence Interval for $\mu_{Y|x}$: We want an interval estimate of the average value of $Y$ given a value of $x$; that is, on $\mu_{Y|x}$. Again, the general form of a CI:

$$\text{pt. est.} \pm RF^*SE$$

point estimate:

t-value:

standard error of $\hat{y}$:

$$S\sqrt{\frac{1}{n} + \frac{(x - \bar{x})^2}{(n - 1)*Var(X)}}$$

Thus, the formula for a CI on the average value of $Y$ given $x$ is

$$\hat{y} \pm t_{(n-2)} \cdot S \cdot \sqrt{\frac{1}{n} + \frac{(x - \bar{x})^2}{(n - 1)*Var(X)}}$$

- Example 2: Find a 95% CI on the average systolic blood pressure given the person exercises strenuously for forty-five minutes a day, four days a week.

  What is $x$?

  Answer:

- We input this value of $x$ into our data in JMP:

  Leave the $y$ blank. JMP will put the period there for you.

| 19 | 4.5 | 117 |
|----|-----|-----|
| 20 | 7.5 | 102 |
| 21 | 3 | . |
| | | |

Under the *linear fit* option, choose _____ .

Output:

| 19 | 4.5 | 117 | 116.2720055 | 121.3480996 |
|----|-----|-----|-------------|-------------|
| 20 | 7.5 | 102 | 98.60171454 | 104.994006 |
| 21 | 3 | . | 127.20459974 | 130.43629445 |
| | | | | |

Thus, the 95% CI for $\mu_{Y|x}$ is:

Interpretation:

We _____, with 95% confidence, that _____

who works out strenuously for _____ a week will have an

_____ systolic blood pressure between _____ and

_____, on average.

- A Prediction Interval, PI, for the value of $y$ given $x$ (*not* the average of $y$, the predicted value of $y$) is found using the formula:

$$\hat{y} \pm t_{\alpha/2} S \sqrt{1 + \frac{1}{n} + \frac{(x - \bar{x})^2}{(n-1)*Var(X)}}$$

Example 3: Suppose we want a 95% prediction interval for the systolic blood pressure for a specific person who works out for forty-five minutes a day, four days a week.

JMP will do this for us.

Under the *linear fit* option, choose _____.

JMP Output:

| | | | | | |
|---|---|---|---|---|---|
| 7.5 | 102 | 98.60171454 | 104.994006 | 94.537745637 | 109.0579749 |
| 3 | • | 127.20459974 | 130.43629445 | 122.10443045 | 135.53646374 |

Interpretation:

We _____, with 95% confidence, that this _____,

who works out strenuously for _____ hours a week will have a systolic

blood pressure between _____ and _____ mmhg,

on average.

Note that a confidence interval gives the average systolic blood pressure expected for *all* subjects who exercise 45 minutes per day, four days a week, whereas the prediction interval gives the average systolic blood pressure expected for a *specific* subject who exercises forty-five minutes a day, four days a week.

The prediction interval is _____ than the confidence interval, this is because one value will vary more (i.e., have more possible values) than several values will.

# Check Your Understanding

A pharmaceutical company has developed a new drug for patients with Wolfe-Parkinson-White syndrome designed to control paroxysmal supraventricular tachycardia (PSVT). Patients with PSVT have a heart rate of more than 100 beats per minute. (BPM) Suppose you are interested in modeling the relationship between the decrease in pulse rate in BPM and dosage in cc. Different doses were administered to 16 randomly selected patients and 20 minutes later the decrease in each patient's pulse rate was recorded. The data collected is given below.

| Dosage | 2 | 4 | 1.5 | 1 | 3 | 3.5 | 2.5 | 3 | 3.5 | 1.5 | 2 | 2 | 3.75 | 3.5 | 1.75 | 1.25 |
|--------|---|---|-----|---|---|-----|-----|---|-----|-----|---|---|------|-----|------|------|
| Drop in HR | 14 | 24 | 11 | 6 | 18 | 22 | 17 | 15 | 22 | 9 | 16 | 14 | 23 | 25 | 12 | 12 |

    **a.** Find and interpret a 99% confidence interval for the slope of the regression line.

    **b.** Find and interpret a 99% confidence interval for the average drop in pulse rate for a dosage level of 2.75 cc.

    **c.** Find and interpret a 99% prediction interval for a dosage level of 2.75 cc.

# ➡ Section 5: Correlation Analysis:

Used for describing the strength of the *linear* relationship between the response and independent variables.

There are two values we can calculate to describe the strength of the linear relationship between two variables.

    **1.** The Correlation Coefficient

    **2.** The Coefficient of Determination

- *The Correlation Coefficient*

   The correlation coefficient measures the strength of the linear relationship between two variables. The correlation coefficient is scale less and denoted by the symbol, *r*.

- The formula for finding *r*:
$$r = \frac{SS_{xy}}{\sqrt{SS_{xy} SS_{xx}}}.$$

- The following plots illustrate various values of *r*. The value of *r* is written below each plot.

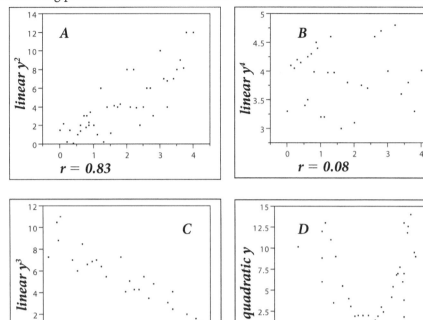

- Which plot represents the strongest linear relationship?

  Answer:

- Which plot represents no relationship?

  Answer:

- Is there a strong relationship being displayed in plot D?

  Answer:

- Is it linear?

  Answer:

- What is the relationship between $r$ and $b_1$?

  Answer:


- The strength of the linear relationship is represented by how far $r$ is from 0:

  - _____ $\leq r \leq$ _____

  - If $r =$ _____ $\rightarrow$

  - If $r =$ _____ $\rightarrow$ perfect linear relationship, which means graphically:

    Answer:


- The sign of $r$ relates information about the type of linear relationship between the two variables:

  If the value of $r$ is positive $\rightarrow$ positive or direct linear relationship.
  Thus _high_ values of one variable are paired with _high_ values of the other

  If the value of $r$ is negative $\rightarrow$ negative or indirect relationship
  Thus _high_ values of one variable are paired with _low_ values of the other

- The value of $r$ is easy to interpret when it is close to zero or positive or negative one, but for "middle" values of $r$, it can be subjective to interpret. In other words, is it a somewhat strong, fairly strong, somewhat weak, fairly weak, etc., linear relationship?

- _The Coefficient of Determination_:

  The coefficient of determination describes the percentage of the variation in the dependent variable that can be accounted for by the linear relationship between $x$ and $y$ and is denoted by $r^2$.

  - The formula for finding $r^2$:  $r^2 = \dfrac{SSR}{SS_{yy}} = 1 - \dfrac{SSE}{SS_{yy}}$.

  - In simple linear regression, $r^2$ is simply the square of $r$.

  - Thus, _____ $\leq r^2 \leq$ _____.

  - There are two ways to interpret $r^2$: Suppose $r^2 = 0.9453$.

    Variance Interpretation: Approximately 94.53% of the variation in $y$ can be explained by the linear relationship between $x$ and $y$.

    Error Interpretation: We will reduce our prediction error using the linear regression equation, $\hat{y}$, to predict $y$ versus using $\bar{y}$ by 94.53%.

- The following tables and plots illustrate the relationship between the slope, correlation, and the coefficient of determination.

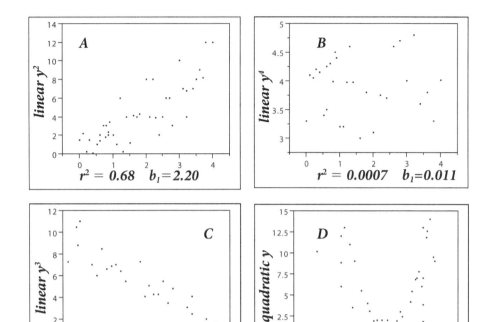

- Example of finding $r$ and $r^2$

  - For our exercise vs blood pressure example: We get $r^2$ from JMP. We can calculate $r$, by taking the square root of $r^2$. We know that the sign of $r$ has to match the sign of $b_1$. So, we have

$r^2 = 0.934533$

Variance Interpretation: Approximately 93.45% of the variation in _____

can be explained by the _____ relationship between _____

and _____ .

Error Interpretation: We will _____ the _____

_____ by approximately _____ using

_____ to estimate _____ versus using

_____ to estimate _____ .

Thus, $r = \sqrt{0.934533}$ = _____ . ($r$ is _____ ,

since $b_1$ is _____ ),

- The correlation coefficient, $r$, and the coefficient of determination, $r^2$, describe the strength of the linear relationship, but does the slope, $b_1$, tell us anything about the strength of the linear relationship?

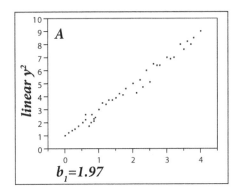 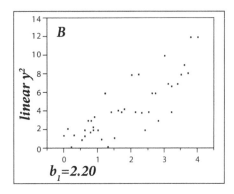

- Which graph represents the variables that have the strongest linear relationship?

  Answer:



- Which regression line has the largest slope?

  Answer:



- So, does the size of the slope have anything to do with the strength of the linear relationship?

  Answer:



- Some comments about correlation:

  - Always, always, always, look at the scatter plot! Why?

    Answer:



  - There is more than one way to measure correlation. The *Pearson product moment correlation coefficient* is the measure we are using and is the most widely used one.

  - Correlation is a dimensionless measure of the strength of the linear relationship. That is, it does not matter what the unit of measurement is for either variable.

  - Correlation measures the strength of the *linear* relationship between variables.

  - *Just because two variables have a strong correlation does NOT mean that a cause-effect relationship exists.* If two variables are highly correlated, it *could* mean that the independent variable is causing the dependent variable to occur, but it does not, on its own, imply that the independent variable is causing the dependent variable to occur. There could be another variable that is causing the independent and dependent variable to behave in the same way.

    For example: Ice cream consumption has a positive correlation with drowning rates. Does that mean eating ice cream causes drowning?

# Check Your Understanding

A pharmaceutical company has developed a new drug for patients with Wolfe-Parkinson-White syndrome designed to control paroxysmal supraventricular tachycardia (PSVT). Patients with PSVT have a heart rate of more than 100 beats per minute. (BPM) Suppose you are interested in modeling the relationship between the decrease in pulse rate in BPM and dosage in cc. Different doses were administered to 16 randomly selected patients and 20 minutes later the decrease in each patient's pulse rate was recorded. The data collected is given below.

| Dosage | 2 | 4 | 1.5 | 1 | 3 | 3.5 | 2.5 | 3 | 3.5 | 1.5 | 2 | 2 | 3.75 | 3.5 | 1.75 | 1.25 |
|--------|---|---|-----|---|---|-----|-----|---|-----|-----|---|---|------|-----|------|------|
| Drop in HR | 14 | 24 | 11 | 6 | 18 | 22 | 17 | 15 | 22 | 9 | 16 | 14 | 23 | 25 | 12 | 12 |

    **a.** Find the correlation coefficient. Is there a strong linear relationship? Is it a direct or an indirect?

    **b.** Find the coefficient of determination.

    **c.** Interpret the coefficient of determination using the variance interpretation.

    **d.** Interpret the coefficient of determination using the error interpretation.

# ➡ Section 6: Residual Analysis:

Is the relationship between amount of time in strenuous exercise and average systolic blood pressure perfectly linear? No! Is the linear model close enough to reality to use in order to predict the average systolic blood pressure? Hopefully so!!

We know the stronger the linear relationship; the better the model will be in predicting the average value of $y$. In addition to verifying that there is a linear relationship between the variables, there are other assumptions that must be true in order for the model to be valid.

Residual analysis is used to determine whether the linear model is appropriate.

- Recall, the population model for simple linear regression: $y = \beta_0 + \beta_1 x + \varepsilon$

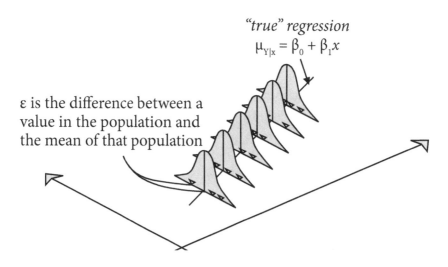

*"true" regression*
$$\mu_{Y|x} = \beta_0 + \beta_1 x$$

$\varepsilon$ is the difference between a value in the population and the mean of that population

*From Dr. Seaman's STA 3381 notes.*

- **Simple Linear Regression Assumptions**: The assumptions on the errors, ε:

  1. The relationship between the independent variable, $x$, and the averages of the dependent variable, $y$, given $x$ is <u>linear</u>; i.e., $E(Y|x) = \beta_0 + \beta_1 x$.

  2. The errors are independent.

  3. The errors have constant variance (homoscedasticity) and zero mean for all values of $x$

  4. The errors are normally distributed.

  We can verify assumptions 1 – 3 by creating a residual plot. We can verify assumption 4 by looking at a normal quantile plot of the residuals. Specific properties of the residuals are what allow us to check the assumptions of linear regression by looking at the residuals.

  Recall, each value has a residual which is defined as $res_i = y_i - \hat{y}_i$.

- Properties of residuals

  1. The average of the residuals is always zero. Why?

     Answer:

  2. If assumptions 1 – 3 are valid, then the residuals graphed on a residual plot should be randomly distributed about zero. In other words, there will not be a pattern in the residuals. If there is a pattern, the type of pattern can tell us what assumption is violated. (Technically, the residuals are not independent. However, except for small data sets and residuals for extreme data points, the correlation among residuals is very small and can be ignored.)

  3. If assumption 4 on the last page is valid, then the residuals should be closely clustered along the line on the QQ-plot.

- So, to check our assumptions, we need to do the following:

  1. Look at the residual plot.

  2. Look at the QQ-Plot of the residuals.

  3. Check for any outliers.

- Residual plots: Plot the residuals vs. the predicted values and look for patterns. The graphs below illustrate typical patterns.

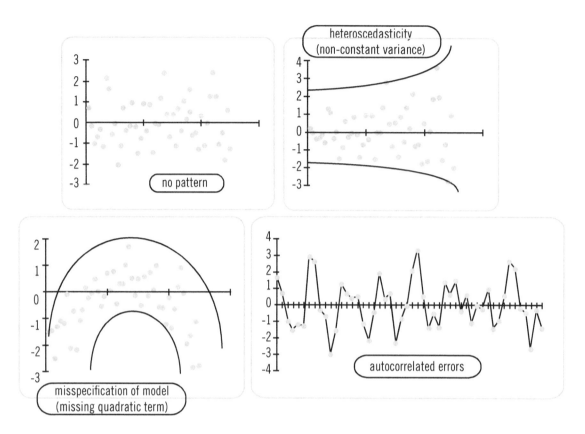

*From Dr. Seaman's STA 3381 notes.*

- **Step 1: The Residual Plot:**

  To get the residual plot, under the linear fit option, choose *plot residual*.

  The residual plot for our blood pressure data:

  Are any of assumptions 1–3 violated?

***Diagnostics Plots***
***Residual by Predicted Plot***

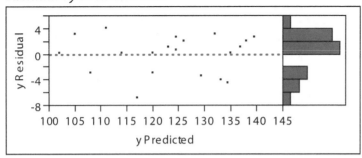

- **Step 2:** The QQ-Plot of the residuals

We look at residuals to see if there is any evidence that the normal error assumption has been violated. We do this with the normal probability plot provided by JMP. Recall, to assume the residuals are normally distributed; the residuals need to be closely clustered to the line.

A Normal Quantile Plot of the residuals is given when the plot residual request is chosen.

Here is the Normal Quantile plot of the residuals:

Can we assume the residuals are normally distributed?

***Diagnostics Plots***
***Residual Normal Quantile Plot***

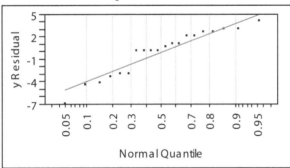

- **Step 3:** Checking for Outliers

  If the residuals are normally distributed, then by the Empirical Rule, we know that about 99.7% of the residuals will fall within three standard deviations of the mean. Thus, since the mean of the residuals equals zero, any residual beyond $\pm 3S = \pm 3(\text{Root MSE})$ is an outlier. Hence, the data value corresponding to that residual is an outlier.

  To determine if there are any outliers, we look at the list of residuals provided by JMP. We need to determine if there are any residuals bigger in magnitude than $\pm 3(\text{Root MSE})$.

  So for our strenuous exercise vs. systolic blood pressure example:

  $\pm 3(\text{Root MSE}) = \pm 3(\underline{\hspace{4cm}}) = \underline{\hspace{4cm}}$.

  Are there any outliers?

**The Residuals:**

| | | | |
|---|---|---|---|
| 1.192877243 | 2.1820625523 | 0.1870818469 | 1.1845721996 |
| 0.1745336103 | −4.345064846 | 0.6833173759 | 0.1921011416 |
| 2.1720239629 | 3.1770432576 | −6.810408506 | −2.802879564 |
| −4.023960601 | 2.6707691393 | 4.1946107889 | −2.812918153 |
| −3.300848639 | 2.6833173759 | 3.1996300836 | 0.2021397309 |

Outliers should always be investigated and never simply deleted! If an outlier is not an error, then that outlier might be trying to provide us with information about some phenomenon occurring. If you do omit the outliers and refit the model after removing the outliers, you should always report the fact that you have done so.

- Some other variables in regression to be aware of:

  - Influential Observations: An influential observation is an observation that, although not an outlier, does influence the regression line, which in turn influences the predictions found.

  - How would the observation, $x$, influence the following regression line?

    Answer:

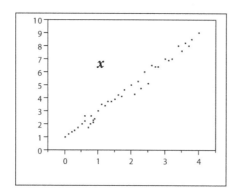

An influential observation can also have an effect on the correlation coefficient. What is that effect?

Answer:

- Lurking Variables: A lurking variable is a variable that you have not considered in your analysis but which does influence the dependent variable in a regression. Lurking variables also affect correlation. A lurking variable has the same effect on the regression model as an extraneous variable has on an experiment.

  Consider the following example courtesy of Dr. John Seaman:

  Suppose you work for a retail concern and you have studied the relationship between your advertising expenditures and sales. You found it to be strong and positive and conclude that the recent increases in advertising outlays have paid off. In reality, your only significant competitor has been hampered by nearby road construction restricting customer access to his establishment. "Ease of customer access" was a lurking variable in this case.

# Check Your Understanding

A pharmaceutical company has developed a new drug for patients with Wolfe-Parkinson-White syndrome designed to control paroxysmal supraventricular tachycardia (PSVT). Patients with PSVT have a heart rate of more than 100 beats per minute. (BPM) Suppose you are interested in modeling the relationship between the decrease in pulse rate in BPM and dosage in cc. Different doses were administered to 16 randomly selected patients and 20 minutes later the decrease in each patient's pulse rate was recorded. The data collected is given below.

| Dosage | 2 | 4 | 1.5 | 1 | 3 | 3.5 | 2.5 | 3 | 3.5 | 1.5 | 2 | 2 | 3.75 | 3.5 | 1.75 | 1.25 |
|--------|---|---|-----|---|---|-----|-----|---|-----|-----|---|---|------|-----|------|------|
| Drop in HR | 14 | 24 | 11 | 6 | 18 | 22 | 17 | 15 | 22 | 9 | 16 | 14 | 23 | 25 | 12 | 12 |

Perform residual analysis to verify the assumptions of the model.

# ➡ Section 7: Linear Regression and Correlation Analysis Overview:

In Simple Linear Regression, the goal is to build a model that represents the relationship inherent between the two variables. The model can then be used for describing the relationship between the two variables and predicting.

The $X$ variable is called the _____ or _____ variable.

The $Y$ variable is called the _____ or _____ variable.

The variable we desire to predict or estimate is the _____ variable.

The line of best fit (or the sample regression line) is: _____

Where the slope, _____ , represents:

And the $y$-intercept, _____ , represents:

The slope _____ has an interpretation, but the $y$-intercept will not have an

interpretation if _____ is done.

To use the model to estimate, we simply evaluate the model at that point. This estimation found

represents the _____ value of $y$ for _____ $x$ at that level.

A CI for $\beta_1$:

Interpretation: "I am _____% confident that if we _____ the (value of x) by _____ unit, then the _____ (value of y) will (increase or decrease ) between _____ and _____, on average."

A CI for $\mu_{y|x}$:

Interpretation: "I am _____% confident that the _____ (value of y) is between _____ and _____ anytime X equals _____, on average."

A PI:

Interpretation: "I am _____% confident that the (value of y) is between _____ and _____ for a specific (value of x) at the _____ level."

Recall, the CI represents the _____ value of y for _____ x at that level, whereas a PI represents the value of y for _____ x at that level.

The correlation coefficient, denoted by the symbol _____, represents the strength of the _____ relationship between the variables.

The values of the correlation coefficient are _____.

If the correlation coefficient is near zero that implies that there is

_____ relationship between the variables.

If the correlation coefficient is near positive or negative one, then that implies that there is a

_____. relationship between the variables.

The coefficient of determination, denoted by the symbol _____, represents the

strength of the _____ relationship between the variables.

The values of the coefficient of determination are _____.

There are two ways to interpret the coefficient of determination:

1. The percentage of the variation in _____ that can be explained by the _____ relationship between $x$ and $y$.

2. The percentage that we reduce our _____ using the linear regression equation to predict _____ versus using _____.

The assumptions of the regression model:

1. The relationship between the independent variable $x$ and the averages of the dependent variable, $y$ is

_____ .

2. The errors are _____ .

3. The errors have _____ _____ and

_____ mean for all values of $x$

4. The errors are _____ distributed.

We verify the assumptions by doing the following three steps:

1.

2.

3.

# Chapter 10 Homework:

## Section 2

1. An insurance salesman would like a model to predict what a healthy person's monthly health insurance premium would be given his or her age. The data listed below gives the monthly premium and age for 28 randomly selected healthy adults.

| Age | Premium | | Age | Premium |
|-----|---------|---|-----|---------|
| 22 | 154 | | 39 | 169 |
| 23 | 153 | | 38 | 170 |
| 25 | 152 | | 40 | 172 |
| 26 | 153 | | 41 | 170 |
| 29 | 156 | | 54 | 221 |
| 31 | 158 | | 56 | 225 |
| 42 | 182 | | 52 | 220 |
| 43 | 180 | | 59 | 232 |
| 46 | 179 | | 57 | 230 |
| 52 | 205 | | 58 | 240 |
| 57 | 210 | | 60 | 249 |
| 58 | 215 | | 48 | 200 |
| 32 | 161 | | 47 | 180 |
| 35 | 160 | | 50 | 191 |

   a. Which variable is the independent variable?

   b. Which variable is the dependent variable?

   c. Is a linear model appropriate? If so, justify. If not, what type of model, if any, would be most appropriate?

2. Does gaining weight increase systolic blood pressure for adult females? And, if so, what is the relationship? This was a question of interest for researchers. A random sample of twenty female adults was obtained. The data below is the weight measured in pounds and systolic blood pressure measured in mmHg.

| Weight | Blood Pressure | | Weight | Blood Pressure |
|---|---|---|---|---|
| 180 | 138 | | 145 | 120 |
| 175 | 135 | | 156 | 125 |
| 195 | 140 | | 135 | 110 |
| 171 | 129 | | 140 | 115 |
| 160 | 126 | | 115 | 108 |
| 158 | 128 | | 149 | 124 |
| 162 | 130 | | 135 | 114 |
| 150 | 135 | | 130 | 105 |
| 194 | 142 | | 130 | 117 |
| 150 | 127 | | 120 | 102 |

a. Which variable is the independent variable?

b. Which variable is the dependent variable?

c. Is a linear model appropriate? If so, justify. If not, what type of model, if any, would be most appropriate?

**3.** Is a college student's freshman GPA a good predictor for the student's GPA at graduation? To investigate this question, thirty students were randomly selected. The data below represents their GPA at the end of their freshman year and at graduation.

| Freshman GPA | Graduation GPA | | Freshman GPA | Graduation GPA | | Freshman GPA | Graduation GPA |
|---|---|---|---|---|---|---|---|
| 2.32 | 2.49 | | 3.86 | 3.38 | | 3.12 | 3 |
| 2.36 | 3.15 | | 2.97 | 3.05 | | 1.28 | 2.64 |
| 1.98 | 3.05 | | 1.88 | 3.26 | | 1.79 | 2.9 |
| 0.98 | 2.28 | | 1.67 | 2.78 | | 1.45 | 2.7 |
| 3.04 | 3.25 | | 1.54 | 1.98 | | 3.9 | 3.85 |
| 3.22 | 2.88 | | 2.88 | 3.28 | | 4 | 3.9 |
| 2.50 | 3.08 | | 3.05 | 2.98 | | 2.88 | 3.02 |
| 2.08 | 2.28 | | 3.17 | 2.88 | | 1.42 | 2.98 |
| 3.05 | 2.38 | | 3.85 | 3.05 | | 3.8 | 2.2 |
| 3.48 | 2.68 | | 3.92 | 3.28 | | 2.5 | 2.9 |

**a.** Which variable is the independent variable?

**b.** Which variable is the dependent variable?

**c.** Is a linear model appropriate? If so, justify. If not, what type of model, if any, would be most appropriate?

4. A question of interest is whether increasing the amount of exercise will decrease blood glucose levels for pre-diabetic adults and if so, in what fashion. A random sample of twenty pre-diabetic adults was obtained. The data below gives the amount of time spent exercising per week (in hours) and the person's blood glucose level (measured in mg/dL).

| Exercise Time | Glucose Levels | | Exercise Time | Glucose Level |
|---|---|---|---|---|
| 1.67 | 155 | | 4.5 | 105 |
| 2 | 150 | | 3.75 | 125 |
| 1.5 | 156 | | 5 | 95 |
| 2.3 | 149 | | 6 | 98 |
| 2.92 | 151 | | 7 | 100 |
| 3.5 | 129 | | 4 | 118 |
| 2.08 | 145 | | 5.5 | 114 |
| 2.5 | 135 | | 6.5 | 98 |
| 1.25 | 154 | | 4.5 | 107 |
| 3.75 | 127 | | 7.5 | 80 |

   a. Which variable is the independent variable?

   b. Which variable is the dependent variable?

   c. Is a linear model appropriate? If so, justify. If not, what type of model, if any, would be most appropriate?

# Section 3

5. Answer the following questions about $\beta_0$:

   1. What does the symbol $\beta_0$ represent?

   2. Does $\beta_0$ always have an interpretation? If not, explain when it does not have an interpretation.

   3. If it does have an interpretation, what is the general interpretation of $\beta_0$?

   4. What is the point estimate for $\beta_0$?

6. Answer the following questions about $\beta_1$:

   a. What does the symbol $\beta_1$ represent?

   b. Does $\beta_1$ always have an interpretation? If not, explain when it does not have an interpretation.

   c. If it does have an interpretation, what is the general interpretation of $\beta_1$?

   d. What is the point estimate for $\beta_1$?

7. Answer the following questions about $\mu_{Y|x}$:

   a. What does the symbol $\mu_{Y|x}$ represent?

   b. What is the point estimate for $\mu_{Y|x}$?

8. Consider again the data describing the relationship between weight and blood pressure described in problem 2. Use the data in problem 2 to answer the following questions.

   a. Find the regression line.

   b. Interpret, if possible, the coefficients of the regression line.

   c. If possible, estimate the average systolic blood pressure for an adult who weighs 165 pounds.

   d. If possible, estimate the average systolic blood pressure for an adult who weighs 295 pounds.

9. Consider again the data describing the relationship between exercise and glucose levels described in problem 4. Use the data in problem 4 to answer the following questions.

   a. Find the regression line.

   b. Interpret, if possible, the coefficients of the regression line.

   c. If possible, estimate the average blood glucose level for an adult who exercises two hours a day, every day of the week.

   d. If possible, estimate the average blood glucose level for an adult who exercises one hour a day, three days a week.

# Section 4

10. For each of the following confidence intervals for $\beta_1$, (i) determine whether there is a linear relationship between the two variables and (ii) if there is a linear relationship between the two variables determine whether $y$ is increasing or decreasing as $x$ increases.

    a. (−3.24, −1.23)          b. (−5, 2)          c. (2.86, 5.92)

**11.** Once again consider the data describing the relationship between weight and blood pressure described in problem 2. Use the data in problem 2 to answer the following questions.

    **a.** Find *and* interpret a 95% confidence interval for the population slope.

    **b.** Find *and* interpret a 95% confidence interval for the average systolic blood pressure for female adults that weigh 158 pounds.

    **c.** Find *and* interpret a 95% prediction interval for a female adult who weighs 158 pounds.

    **d.** Explicitly explain the difference between the CI found in part b and the PI found in part c.

**12.** Consider once again the data describing the relationship between exercise and glucose levels described in problem 4. Use the data in problem 4 to answer the following questions.

    **a.** Find *and* interpret a 99% confidence interval for the population slope.

    **b.** Find *and* interpret a 99% confidence interval for the average blood glucose level for adults that exercise an hour a day, three days a week.

    **c.** Find *and* interpret a 99% prediction interval for an adult that exercises an hour a day, three days a week.

    **d.** Explicitly explain the difference between the CI found in part b and the PI found in part c.

**13.** For the following two intervals, determine which one is the CI and which one is the PI. Note that all other information was kept the same when constructing the intervals.

    Interval 1: (145, 172)        Interval 2: (154, 162)

# Section 5

**14.** What symbol is used to represent the correlation coefficient?

**15.** What symbol is used to represent the coefficient of determination?

**16.** Suppose you are given the four coefficient of determination values below. Determine which one represents the two variables that are most linearly related.

    **a.** 0.89        **b.** −0.92        **c.** 0.01        **d.** 0.54

**17.** Suppose you are given the four correlation coefficient values below. Determine which one represents the two variables that are most linearly related.

    **a.** 0.89           **b.** −0.92           **c.** 0.01           **d.** 0.54

**18.** For each of the following correlation coefficient values, determine whether you can conclude that the two variables are linearly related and if so, if the value of $y$ increases or decreases as $x$ increases.

    **a.** −0.95          **b.** 0.001          **c.** 0.92          **d.** −0.00005

**19.** Once again consider the data describing the relationship between weight and blood pressure described in problem #2. Use the data in problem #2 to answer the following questions.

    **a.** Find *and* interpret the correlation coefficient.

    **b.** Find the coefficient of determination.

    **c.** Interpret the coefficient of determination using the "variance" interpretation.

    **d.** Interpret the coefficient of determination using the "error" interpretation.

**20.** Consider once again the data describing the relationship between exercise and glucose levels described in problem 4. Use the data in problem 4 to answer the following questions.

    **a.** Find *and* interpret the correlation coefficient.

    **b.** Find the coefficient of determination.

    **c.** Interpret the coefficient of determination using the "variance" interpretation.

    **d.** Interpret the coefficient of determination using the "error" interpretation.

# Section 6

**21.** State the assumptions of the linear regression model.

**22.** Once again consider the data describing the relationship between weight and blood pressure described in problem 2. Use the data in problem 2 to answer the following questions.

    **a.** Perform the residual analysis to verify the assumptions of the model. If any assumptions are violated, clearly state which assumption(s) are violated.

    **b.** Determine if there are any outliers. If there are outliers, state which data value is an outlier.

**23.** Consider once again the data describing the relationship between exercise and glucose levels described in problem 4. Use the data in problem 4 to answer the following questions.

    **a.** Perform the residual analysis to verify the assumptions of the model. If any assumptions are violated, clearly state which assumption(s) are violated.

    **b.** Determine if there are any outliers. If there are outliers, state which data value is an outlier.

**24.** Define the following terms:

    **a.** A lurking variable.

    **b.** An influential observation.

area to left!

RMULA/TABLE CARD FOR WEISS'S *INTRODUCTORY STATISTICS, EIGHTH EDITION*

Larry R. Griffey

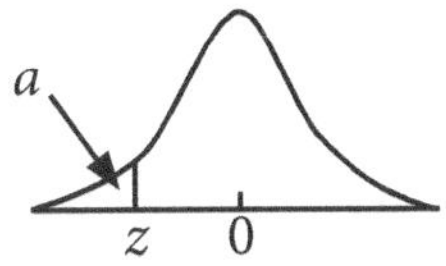


TABLE II Areas under the standard normal curve

| Second decimal place in z | | | | | | | | | | |
|---|---|---|---|---|---|---|---|---|---|---|
| *0.09* | *0.08* | *0.07* | *0.06* | *0.05* | *0.04* | *0.03* | *0.02* | *0.01* | *0.00* | *z* |
| | | | | | | | | | 0.0000* | −3.9 |
| 0.0001 | 0.0001 | 0.0001 | 0.0001 | 0.0001 | 0.0001 | 0.0001 | 0.0001 | 0.0001 | 0.0001 | −3.8 |
| 0.0001 | 0.0001 | 0.0001 | 0.0001 | 0.0001 | 0.0001 | 0.0001 | 0.0001 | 0.0001 | 0.0001 | −3.7 |
| 0.0001 | 0.0001 | 0.0001 | 0.0001 | 0.0001 | 0.0001 | 0.0001 | 0.0001 | 0.0002 | 0.0002 | −3.6 |
| 0.0002 | 0.0002 | 0.0002 | 0.0002 | 0.0002 | 0.0002 | 0.0002 | 0.0002 | 0.0002 | 0.0002 | −3.5 |
| 0.0002 | 0.0003 | 0.0003 | 0.0003 | 0.0003 | 0.0003 | 0.0003 | 0.0003 | 0.0003 | 0.0003 | −3.4 |
| 0.0003 | 0.0004 | 0.0004 | 0.0004 | 0.0004 | 0.0004 | 0.0004 | 0.0005 | 0.0005 | 0.0005 | −3.3 |
| 0.0005 | 0.0005 | 0.0005 | 0.0006 | 0.0006 | 0.0006 | 0.0006 | 0.0006 | 0.0007 | 0.0007 | −3.2 |
| 0.0007 | 0.0007 | 0.0008 | 0.0008 | 0.0008 | 0.0008 | 0.0009 | 0.0009 | 0.0009 | 0.0010 | −3.1 |
| 0.0010 | 0.0010 | 0.0011 | 0.0011 | 0.0011 | 0.0012 | 0.0012 | 0.0013 | 0.0013 | 0.0013 | −3.0 |
| 0.0014 | 0.0014 | 0.0015 | 0.0015 | 0.0016 | 0.0016 | 0.0017 | 0.0018 | 0.0018 | 0.0019 | −2.9 |
| 0.0019 | 0.0020 | 0.0021 | 0.0021 | 0.0022 | 0.0023 | 0.0023 | 0.0024 | 0.0025 | 0.0026 | −2.8 |
| 0.0026 | 0.0027 | 0.0028 | 0.0029 | 0.0030 | 0.0031 | 0.0032 | 0.0033 | 0.0034 | 0.0035 | −2.7 |
| 0.0036 | 0.0037 | 0.0038 | 0.0039 | 0.0040 | 0.0041 | 0.0043 | 0.0044 | 0.0045 | 0.0047 | −2.6 |
| 0.0048 | 0.0049 | 0.0051 | 0.0052 | 0.0054 | 0.0055 | 0.0057 | 0.0059 | 0.0060 | 0.0062 | −2.5 |
| 0.0064 | 0.0066 | 0.0068 | 0.0069 | 0.0071 | 0.0073 | 0.0075 | 0.0078 | 0.0080 | 0.0082 | −2.4 |
| 0.0084 | 0.0087 | 0.0089 | 0.0091 | 0.0094 | 0.0096 | 0.0099 | 0.0102 | 0.0104 | 0.0107 | −2.3 |
| 0.0110 | 0.0113 | 0.0116 | 0.0119 | 0.0122 | 0.0125 | 0.0129 | 0.0132 | 0.0136 | 0.0139 | −2.2 |
| 0.0143 | 0.0146 | 0.0150 | 0.0154 | 0.0158 | 0.0162 | 0.0166 | 0.0170 | 0.0174 | 0.0179 | −2.1 |
| 0.0183 | 0.0188 | 0.0192 | 0.0197 | 0.0202 | 0.0207 | 0.0212 | 0.0217 | 0.0222 | 0.0228 | −2.0 |
| 0.0233 | 0.0239 | 0.0244 | 0.0250 | 0.0256 | 0.0262 | 0.0268 | 0.0274 | 0.0281 | 0.0287 | −1.9 |
| 0.0294 | 0.0301 | 0.0307 | 0.0314 | 0.0322 | 0.0329 | 0.0336 | 0.0344 | 0.0351 | 0.0359 | −1.8 |
| 0.0367 | 0.0375 | 0.0384 | 0.0392 | 0.0401 | 0.0409 | 0.0418 | 0.0427 | 0.0436 | 0.0446 | −1.7 |
| 0.0455 | 0.0465 | 0.0475 | 0.0485 | 0.0495 | 0.0505 | 0.0516 | 0.0526 | 0.0537 | 0.0548 | −1.6 |
| 0.0559 | 0.0571 | 0.0582 | 0.0594 | 0.0606 | 0.0618 | 0.0630 | 0.0643 | 0.0655 | 0.0668 | −1.5 |
| 0.0681 | 0.0694 | 0.0708 | 0.0721 | 0.0735 | 0.0749 | 0.0764 | 0.0778 | 0.0793 | 0.0808 | −1.4 |
| 0.0823 | 0.0838 | 0.0853 | 0.0869 | 0.0885 | 0.0901 | 0.0918 | 0.0934 | 0.0951 | 0.0968 | −1.3 |
| 0.0985 | 0.1003 | 0.1020 | 0.1038 | 0.1056 | 0.1075 | 0.1093 | 0.1112 | 0.1131 | 0.1151 | −1.2 |
| 0.1170 | 0.1190 | 0.1210 | 0.1230 | 0.1251 | 0.1271 | 0.1292 | 0.1314 | 0.1335 | 0.1357 | −1.1 |
| 0.1379 | 0.1401 | 0.1423 | 0.1446 | 0.1469 | 0.1492 | 0.1515 | 0.1539 | 0.1562 | 0.1587 | −1.0 |
| 0.1611 | 0.1635 | 0.1660 | 0.1685 | 0.1711 | 0.1736 | 0.1762 | 0.1788 | 0.1814 | 0.1841 | −0.9 |
| 0.1867 | 0.1894 | 0.1922 | 0.1949 | 0.1977 | 0.2005 | 0.2033 | 0.2061 | 0.2090 | 0.2119 | −0.8 |
| 0.2148 | 0.2177 | 0.2206 | 0.2236 | 0.2266 | 0.2296 | 0.2327 | 0.2358 | 0.2389 | 0.2420 | −0.7 |
| 0.2451 | 0.2483 | 0.2514 | 0.2546 | 0.2578 | 0.2611 | 0.2643 | 0.2676 | 0.2709 | 0.2743 | −0.6 |
| 0.2776 | 0.2810 | 0.2843 | 0.2877 | 0.2912 | 0.2946 | 0.2981 | 0.3015 | 0.3050 | 0.3085 | −0.5 |
| 0.3121 | 0.3156 | 0.3192 | 0.3228 | 0.3264 | 0.3300 | 0.3336 | 0.3372 | 0.3409 | 0.3446 | −0.4 |
| 0.3483 | 0.3520 | 0.3557 | 0.3594 | 0.3632 | 0.3669 | 0.3707 | 0.3745 | 0.3783 | 0.3821 | −0.3 |
| 0.3859 | 0.3897 | 0.3936 | 0.3974 | 0.4013 | 0.4052 | 0.4090 | 0.4129 | 0.4168 | 0.4207 | −0.2 |
| 0.4247 | 0.4286 | 0.4325 | 0.4364 | 0.4404 | 0.4443 | 0.4483 | 0.4522 | 0.4562 | 0.4602 | −0.1 |
| 0.4641 | 0.4681 | 0.4721 | 0.4761 | 0.4801 | 0.4840 | 0.4880 | 0.4920 | 0.4960 | 0.5000 | −0.0 |

* For $z \leq -3.90$. the areas are 0.0000 to four decimal places.

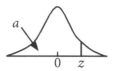

TABLE II    (cont.)    Areas under the standard normal curve

| z | Second decimal place in z | | | | | | | | | |
|---|------|------|------|------|------|------|------|------|------|------|
| | 0.00 | 0.01 | 0.02 | 0.03 | 0.04 | 0.05 | 0.06 | 0.07 | 0.08 | 0.09 |
| 0.0 | 0.5000 | 0.5040 | 0.5080 | 0.5120 | 0.5160 | 0.5199 | 0.5239 | 0.5279 | 0.5319 | 0.5359 |
| 0.1 | 0.5398 | 0.5438 | 0.5478 | 0.5517 | 0.5557 | 0.5596 | 0.5636 | 0.5675 | 0.5714 | 0.5753 |
| 0.2 | 0.5793 | 0.5832 | 0.5871 | 0.5910 | 0.5948 | 0.5987 | 0.6026 | 0.6064 | 0.6103 | 0.6141 |
| 0.3 | 0.6179 | 0.6217 | 0.6255 | 0.6293 | 0.6331 | 0.6368 | 0.6406 | 0.6443 | 0.6480 | 0.6517 |
| 0.4 | 0.6554 | 0.6591 | 0.6628 | 0.6664 | 0.6700 | 0.6736 | 0.6772 | 0.6808 | 0.6844 | 0.6879 |
| 0.5 | 0.6915 | 0.6950 | 0.6985 | 0.7019 | 0.7054 | 0.7088 | 0.7123 | 0.7157 | 0.7190 | 0.7224 |
| 0.6 | 0.7257 | 0.7291 | 0.7324 | 0.7357 | 0.7389 | 0.7422 | 0.7454 | 0.7486 | 0.7517 | 0.7549 |
| 0.7 | 0.7580 | 0.7611 | 0.7642 | 0.7673 | 0.7704 | 0.7734 | 0.7764 | 0.7794 | 0.7823 | 0.7852 |
| 0.8 | 0.7881 | 0.7910 | 0.7939 | 0.7967 | 0.7995 | 0.8023 | 0.8051 | 0.8078 | 0.8106 | 0.8133 |
| 0.9 | 0.8159 | 0.8186 | 0.8212 | 0.8238 | 0.8264 | 0.8289 | 0.8315 | 0.8340 | 0.8365 | 0.8389 |
| 1.0 | 0.8413 | 0.8438 | 0.8461 | 0.8485 | 0.8508 | 0.8531 | 0.8554 | 0.8577 | 0.8599 | 0.8621 |
| 1.1 | 0.8643 | 0.8665 | 0.8686 | 0.8708 | 0.8729 | 0.8749 | 0.8770 | 0.8790 | 0.8810 | 0.8830 |
| 1.2 | 0.8849 | 0.8869 | 0.8888 | 0.8907 | 0.8925 | 0.8944 | 0.8962 | 0.8980 | 0.8997 | 0.9015 |
| 1.3 | 0.9032 | 0.9049 | 0.9066 | 0.9082 | 0.9099 | 0.9115 | 0.9131 | 0.9147 | 0.9162 | 0.9177 |
| 1.4 | 0.9192 | 0.9207 | 0.9222 | 0.9236 | 0.9251 | 0.9265 | 0.9279 | 0.9292 | 0.9306 | 0.9319 |
| 1.5 | 0.9332 | 0.9345 | 0.9357 | 0.9370 | 0.9382 | 0.9394 | 0.9406 | 0.9418 | 0.9429 | 0.9441 |
| 1.6 | 0.9452 | 0.9463 | 0.9474 | 0.9484 | 0.9495 | 0.9505 | 0.9515 | 0.9525 | 0.9535 | 0.9545 |
| 1.7 | 0.9554 | 0.9564 | 0.9573 | 0.9582 | 0.9591 | 0.9599 | 0.9608 | 0.9616 | 0.9625 | 0.9633 |
| 1.8 | 0.9641 | 0.9649 | 0.9656 | 0.9664 | 0.9671 | 0.9678 | 0.9686 | 0.9693 | 0.9699 | 0.9706 |
| 1.9 | 0.9713 | 0.9719 | 0.9726 | 0.9732 | 0.9738 | 0.9744 | 0.9750 | 0.9756 | 0.9761 | 0.9767 |
| 2.0 | 0.9772 | 0.9778 | 0.9783 | 0.9788 | 0.9793 | 0.9798 | 0.9803 | 0.9808 | 0.9812 | 0.9817 |
| 2.1 | 0.9821 | 0.9826 | 0.9830 | 0.9834 | 0.9838 | 0.9842 | 0.9846 | 0.9850 | 0.9854 | 0.9857 |
| 2.2 | 0.9861 | 0.9864 | 0.9868 | 0.9871 | 0.9875 | 0.9878 | 0.9881 | 0.9884 | 0.9887 | 0.9890 |
| 2.3 | 0.9893 | 0.9896 | 0.9898 | 0.9901 | 0.9904 | 0.9906 | 0.9909 | 0.9911 | 0.9913 | 0.9916 |
| 2.4 | 0.9918 | 0.9920 | 0.9922 | 0.9925 | 0.9927 | 0.9929 | 0.9931 | 0.9932 | 0.9934 | 0.9936 |
| 2.5 | 0.9938 | 0.9940 | 0.9941 | 0.9943 | 0.9945 | 0.9946 | 0.9948 | 0.9949 | 0.9951 | 0.9952 |
| 2.6 | 0.9953 | 0.9955 | 0.9956 | 0.9957 | 0.9959 | 0.9960 | 0.9961 | 0.9962 | 0.9963 | 0.9964 |
| 2.7 | 0.9965 | 0.9966 | 0.9967 | 0.9968 | 0.9969 | 0.9970 | 0.9971 | 0.9972 | 0.9973 | 0.9974 |
| 2.8 | 0.9974 | 0.9975 | 0.9976 | 0.9977 | 0.9977 | 0.9978 | 0.9979 | 0.9979 | 0.9980 | 0.9981 |
| 2.9 | 0.9981 | 0.9982 | 0.9982 | 0.9983 | 0.9984 | 0.9984 | 0.9985 | 0.9985 | 0.9986 | 0.9986 |
| 3.0 | 0.9987 | 0.9987 | 0.9987 | 0.9988 | 0.9988 | 0.9989 | 0.9989 | 0.9989 | 0.9990 | 0.9990 |
| 3.1 | 0.9990 | 0.9991 | 0.9991 | 0.9991 | 0.9992 | 0.9992 | 0.9992 | 0.9992 | 0.9993 | 0.9993 |
| 3.2 | 0.9993 | 0.9993 | 0.9994 | 0.9994 | 0.9994 | 0.9994 | 0.9994 | 0.9995 | 0.9995 | 0.9995 |
| 3.3 | 0.9995 | 0.9995 | 0.9995 | 0.9996 | 0.9996 | 0.9996 | 0.9996 | 0.9996 | 0.9996 | 0.9997 |
| 3.4 | 0.9997 | 0.9997 | 0.9997 | 0.9997 | 0.9997 | 0.9997 | 0.9997 | 0.9997 | 0.9997 | 0.9998 |
| 3.5 | 0.9998 | 0.9998 | 0.9998 | 0.9998 | 0.9998 | 0.9998 | 0.9998 | 0.9998 | 0.9998 | 0.9998 |
| 3.6 | 0.9998 | 0.9998 | 0.9999 | 0.9999 | 0.9999 | 0.9999 | 0.9999 | 0.9999 | 0.9999 | 0.9999 |
| 3.7 | 0.9999 | 0.9999 | 0.9999 | 0.9999 | 0.9999 | 0.9999 | 0.9999 | 0.9999 | 0.9999 | 0.9999 |
| 3.8 | 0.9999 | 0.9999 | 0.9999 | 0.9999 | 0.9999 | 0.9999 | 0.9999 | 0.9999 | 0.9999 | 0.9999 |
| 3.9 | 1.0000* | | | | | | | | | |

* For $z \geq 3.90$. the areas are 1.0000 to four decimal places.

Larry R. Griffey

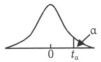

TABLE IV    Values of $t_\alpha$

| df | $t_{0.10}$ | $t_{0.05}$ | $t_{0.025}$ | $t_{0.01}$ | $t_{0.005}$ | df |
|---|---|---|---|---|---|---|
| 1 | 3.078 | 6.314 | 12.706 | 31.821 | 63.657 | 1 |
| 2 | 1.886 | 2.920 | 4.303 | 6.965 | 9.925 | 2 |
| 3 | 1.638 | 2.353 | 3.182 | 4.541 | 5.841 | 3 |
| 4 | 1.533 | 2.132 | 2.776 | 3.747 | 4.604 | 4 |
| 5 | 1.476 | 2.015 | 2.571 | 3.365 | 4.032 | 5 |
| 6 | 1.440 | 1.943 | 2.447 | 3.143 | 3.707 | 6 |
| 7 | 1.415 | 1.895 | 2.365 | 2.998 | 3.499 | 7 |
| 8 | 1.397 | 1.860 | 2.306 | 2.896 | 3.355 | 8 |
| 9 | 1.383 | 1.833 | 2.262 | 2.821 | 3.250 | 9 |
| 10 | 1.372 | 1.812 | 2.228 | 2.764 | 3.169 | 10 |
| 11 | 1.363 | 1.796 | 2.201 | 2.718 | 3.106 | 11 |
| 12 | 1.356 | 1.782 | 2.179 | 2.681 | 3.055 | 12 |
| 13 | 1.350 | 1.771 | 2.160 | 2.650 | 3.012 | 13 |
| 14 | 1.345 | 1.761 | 2.145 | 2.624 | 2.977 | 14 |
| 15 | 1.341 | 1.753 | 2.131 | 2.602 | 2.947 | 15 |
| 16 | 1.337 | 1.746 | 2.120 | 2.583 | 2.921 | 16 |
| 17 | 1.333 | 1.740 | 2.110 | 2.567 | 2.898 | 17 |
| 18 | 1.330 | 1.734 | 2.101 | 2.552 | 2.878 | 18 |
| 19 | 1.328 | 1.729 | 2.093 | 2.539 | 2.861 | 19 |
| 20 | 1.325 | 1.725 | 2.086 | 2.528 | 2.845 | 20 |
| 21 | 1.323 | 1.721 | 2.080 | 2.518 | 2.831 | 21 |
| 22 | 1.321 | 1.717 | 2.074 | 2.508 | 2.819 | 22 |
| 23 | 1.319 | 1.714 | 2.069 | 2.500 | 2.807 | 23 |
| 24 | 1.318 | 1.711 | 2.064 | 2.492 | 2.797 | 24 |
| 25 | 1.316 | 1.708 | 2.060 | 2.485 | 2.787 | 25 |
| 26 | 1.315 | 1.706 | 2.056 | 2.479 | 2.779 | 26 |
| 27 | 1.314 | 1.703 | 2.052 | 2.473 | 2.771 | 27 |
| 28 | 1.313 | 1.701 | 2.048 | 2.467 | 2.763 | 2S |
| 29 | 1.311 | 1.699 | 2.045 | 2.462 | 2.756 | 29 |
| 30 | 1.310 | 1.697 | 2.042 | 2.457 | 2.750 | 30 |
| 31 | 1.309 | 1.696 | 2.040 | 2.453 | 2.744 | 31 |
| 32 | 1.309 | 1.694 | 2.037 | 2.449 | 2.738 | 32 |
| 33 | 1.308 | 1.692 | 2.035 | 2.445 | 2.733 | 33 |
| 34 | 1.307 | 1.691 | 2.032 | 2.441 | 2.728 | 34 |
| 35 | 1.306 | 1.690 | 2.030 | 2.438 | 2.724 | 35 |
| 36 | 1.306 | 1.688 | 2.028 | 2.434 | 2.719 | 36 |
| 37 | 1.305 | 1.687 | 2.026 | 2.431 | 2.715 | 37 |
| 38 | 1.304 | 1.686 | 2.024 | 2.429 | 2.712 | 38 |
| 39 | 1.304 | 1.685 | 2.023 | 2.426 | 2.708 | 39 |
| 40 | 1.303 | 1.684 | 2.021 | 2.423 | 2.704 | 40 |
| 41 | 1.303 | 1.683 | 2.020 | 2.421 | 2.701 | 41 |
| 42 | 1.302 | 1.682 | 2.018 | 2.418 | 2.698 | 42 |
| 43 | 1.302 | 1.681 | 2.017 | 2.416 | 2.695 | 43 |
| 44 | 1.301 | 1.680 | 2.015 | 2.414 | 2.692 | 44 |
| 45 | 1.301 | 1.679 | 2.014 | 2.412 | 2.690 | 45 |
| 46 | 1.300 | 1.679 | 2.013 | 2.410 | 2.687 | 46 |
| 47 | 1.300 | 1.678 | 2.012 | 2.408 | 2.685 | 47 |
| 48 | 1.299 | 1.677 | 2.011 | 20407 | 2.682 | 48 |
| 49 | 1.299 | 1.677 | 2.010 | 2.405 | 2.680 | 49 |

TABLE IV    (cont.)    Values of $t_\alpha$

| df | $t_{0.10}$ | $t_{0.05}$ | $t_{0.025}$ | $t_{0.01}$ | $t_{0.005}$ | df |
|---|---|---|---|---|---|---|
| 50 | 1.299 | 1.676 | 2.009 | 2.403 | 2.678 | 50 |
| 51 | 1.298 | 1.675 | 2.008 | 2.402 | 2.676 | 51 |
| 52 | 1.298 | 1.675 | 2.007 | 2.400 | 2.674 | 52 |
| 53 | 1.298 | 1.674 | 2.006 | 2.399 | 2.672 | 53 |
| 54 | 1.297 | 1.674 | 2.005 | 2.397 | 2.670 | 54 |
| 55 | 1.297 | 1.673 | 2.004 | 2.396 | 2.668 | 55 |
| 56 | 1.297 | 1.673 | 2.003 | 2.395 | 2.667 | 56 |
| 57 | 1.297 | 1.672 | 2.002 | 2.394 | 2.665 | 57 |
| 58 | 1.296 | 1.672 | 2.002 | 2.392 | 2.663 | 58 |
| 59 | 1.296 | 1.671 | 2.001 | 2.391 | 2.662 | 59 |
| 60 | 1.296 | 1.671 | 2.000 | 2.390 | 2.660 | 60 |
| 61 | 1.296 | 1.670 | 2.000 | 2.389 | 2.659 | 61 |
| 62 | 1.295 | 1.670 | 1.999 | 2.388 | 2.657 | 62 |
| 63 | 1.295 | 1.669 | 1.998 | 2.387 | 2.656 | 63 |
| 64 | 1.295 | 1.669 | 1.998 | 2.386 | 2.655 | 64 |
| 65 | 1.295 | 1.669 | 1.997 | 2.385 | 2.654 | 65 |
| 66 | 1.295 | 1.668 | 1.997 | 2.384 | 2.652 | 66 |
| 67 | 1.294 | 1.668 | 1.996 | 2.383 | 2.651 | 67 |
| 68 | 1.294 | 1.668 | 1.995 | 2.382 | 2.650 | 68 |
| 69 | 1.294 | 1.667 | 1.995 | 2.382 | 2.649 | 69 |
| 70 | 1.294 | 1.667 | 1.994 | 2.381 | 2.648 | 70 |
| 71 | 1.294 | 1.667 | 1.994 | 2.380 | 2.647 | 71 |
| 72 | 1.293 | 1.666 | 1.993 | 2.379 | 2.646 | 72 |
| 73 | 1.293 | 1.666 | 1.993 | 2.379 | 2.645 | 73 |
| 74 | 1.293 | 1.666 | 1.993 | 2.378 | 2.644 | 74 |
| 75 | 1.293 | 1.665 | 1.992 | 2.377 | 2.643 | 75 |
| 80 | 1.292 | 1.664 | 1.990 | 2.374 | 2.639 | 80 |
| 85 | 1.292 | 1.663 | 1.988 | 2.371 | 2.635 | 85 |
| 90 | 1.291 | 1.662 | 1.987 | 2.368 | 2.632 | 90 |
| 95 | 1.291 | 1.661 | 1.985 | 2.366 | 2.629 | 95 |
| 100 | 1.290 | 1.660 | 1.984 | 2.364 | 2.626 | 100 |
| 200 | 1.286 | 1.653 | 1.972 | 2.345 | 2.601 | 200 |
| 300 | 1.284 | 1.650 | 1.968 | 2.339 | 2.592 | 300 |
| 400 | 1.284 | 1.649 | 1.966 | 2.336 | 2.588 | 400 |
| 500 | 1.283 | 1.648 | 1.965 | 2.334 | 2.586 | 500 |
| 600 | 1.283 | 1.647 | 1.964 | 2.333 | 2.584 | 600 |
| 700 | 1.283 | 1.647 | 1.963 | 2.332 | 2.583 | 700 |
| 800 | 1.283 | 1.647 | 1.963 | 2.331 | 2.582 | 800 |
| 900 | 1.282 | 1.647 | 1.963 | 2.330 | 2.581 | 900 |
| 1000 | 1.282 | 1.646 | 1.962 | 2.330 | 2.581 | 1000 |
| 2000 | 1.282 | 1.646 | 1.961 | 2.328 | 2.578 | 2000 |

| | | | | |
|---|---|---|---|---|
| 1.282 | 1.645 | 1.960 | 2.326 | 2.576 |
| $z_{0.10}$ | $z_{0.05}$ | $z_{0.025}$ | $z_{0.01}$ | $z_{0.005}$ |

# FORMULA/TABLE CARD FOR WEISS'S *INTRODUCTORY STATISTICS, EIGHTH EDITION*

## Larry R. Griffey

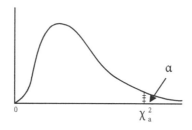

TABLE VII    Values of $\chi^2_{\alpha}$

| $\chi^2_{0.10}$ | $\chi^2_{0.05}$ | $\chi^2_{0.025}$ | $\chi^2_{0.01}$ | $\chi^2_{0.005}$ | df |
|---|---|---|---|---|---|
| 2.706 | 3.841 | 5.024 | 6.635 | 7.879 | 1 |
| 4.605 | 5.991 | 7.378 | 9.210 | 10.597 | 2 |
| 6.251 | 7.815 | 9.348 | 11.345 | 12.838 | 3 |
| 7.779 | 9.488 | 11.143 | 13.277 | 14.860 | 4 |
| 9.236 | 11.070 | 12.833 | 15.086 | 16.750 | 5 |
| 10.645 | 12.592 | 14.449 | 16.812 | 18.548 | 6 |
| 12.011 | 14.061 | 16.013 | 18.475 | 20.278 | 7 |
| 13.362 | 15.507 | 17.535 | 20.090 | 21.955 | 8 |
| 14.684 | 16.919 | 19.023 | 21.666 | 23.589 | 9 |
| 15.987 | 18.307 | 20.483 | 23.209 | 25.188 | 10 |
| 17.275 | 19.675 | 21.920 | 24.725 | 26.757 | 11 |
| 18.549 | 21.026 | 23.337 | 26.217 | 28.300 | 12 |
| 19.812 | 22.362 | 24.736 | 27.688 | 29.819 | 13 |
| 21.064 | 23.685 | 26.119 | 29.141 | 31.319 | 14 |
| 22.307 | 24.996 | 21.488 | 30.578 | 32.801 | 15 |
| 23.542 | 26.296 | 28.845 | 32.000 | 34.267 | 16 |
| 24.769 | 27.587 | 30.191 | 33.409 | 35.718 | 17 |
| 25.989 | 28.869 | 31.526 | 34.805 | 37.156 | 18 |
| 27.204 | 30.143 | 32.852 | 36.191 | 38.582 | 19 |
| 28.412 | 31.410 | 34.170 | 37.566 | 39.997 | 20 |
| 29.615 | 32.611 | 35.419 | 38.932 | 41.401 | 21 |
| 30.813 | 33.924 | 36.781 | 40.290 | 42.796 | 22 |
| 32.007 | 35.172 | 38.076 | 41.638 | 44.181 | 23 |
| 33.196 | 36.415 | 39.364 | 42.980 | 45.559 | 24 |
| 34.382 | 37.653 | 40.647 | 44.314 | 46.928 | 25 |
| 35.563 | 38.885 | 41.923 | 45.642 | 48.290 | 26 |
| 36.741 | 40.113 | 43.195 | 46.963 | 49.645 | 27 |
| 37.916 | 41337 | 44.461 | 48.278 | 50.994 | 28 |
| 39.087 | 42.557 | 45.722 | 49.588 | 52.336 | 29 |
| 40.256 | 43.773 | 46.979 | 50.892 | 53.672 | 30 |
| 51.805 | 55.759 | 59.344 | 63.691 | 66.167 | 40 |
| 63.161 | 61.505 | 71.420 | 76.154 | 79.490 | 50 |
| 74.397 | 19.082 | 83.298 | 88.381 | 91.955 | 60 |
| 85.527 | 90.531 | 95.023 | 100.424 | 104.213 | 70 |
| 96.578 | 101.879 | 106.628 | 112.328 | 116.320 | 80 |
| 107.565 | 113.145 | 118.135 | 124.115 | 128.296 | 90 |
| 118.499 | 124.343 | 129.563 | 135.811 | 140.177 | 100 |

# APPENDIX

**Answers to the *Check Your Understanding* Problems**

## Chapter 2

### Section 3:

1. Suppose the cumulative distribution function for the random variable $X$ is as follows:

| $X$ | 0 | 0.5 | 1 | 1.5 | 2 | 4 |
|---|---|---|---|---|---|---|
| $F(x)$ | 0.47 | 0.62 | 0.73 | a | 0.98 | b |

    **a.** If the $P(X = 1.5) = 0.08$ what is the value of a?

        $a = P(X \le 1.5) = P(X < 1.5) + P(X = 1.5) = P(X \le 1) + P(X = 1.5) = 0.73 + 0.08 = 0.81$

    **b.** What is the value of b?

        $b = P(X \le 4) = F(4) = 1$

    **c.** Find $P(X = 2.5) = 0$ since $X = 2.5$ cannot occur.

        We know it cannot occur since it is not listed in the table.

    **d.** Find the probability that $X$ is at least 1.5 hours.

        $P(X \ge 1.5) = 1 - F(1) = 1 - 0.73 = 0.27$

    **e.** Find $P(0 < X \le 2) = F(2) - F(0) = 0.98 - 0.47 = 0.51$

2. Suppose you play a dice game where if you roll a composite number, you win $5.00. If you roll a one, you win $10. If you roll a prime number, you lose $8. You must pay $2.00 to play. Let $X =$ net winnings.

    **a.** Find the PMF.

| $X$ | −10 | 3 | 8 |
|---|---|---|---|
| $f(x)$ | $\frac{3}{6}$ | $\frac{2}{6}$ | $\frac{1}{6}$ |

    (Typically $\frac{3}{6}$ is reduced to $\frac{1}{2}$ and $\frac{2}{6}$ is reduced to $\frac{1}{3}$)

    **b.** Find the expected value.

        $E(X) = -10\left(\frac{1}{2}\right) + 3\left(\frac{1}{3}\right) + 8\left(\frac{1}{6}\right) = -\frac{16}{6} = -2.67$

    **c.** Interpret the expected value.

        If we play this game over and over again, in the long run, we can expect to lose $2.67 per game.

## Section 4:

Montana Table Company produces custom-made furniture for restaurants, diners and hotels. All of their furniture is handmade with it being known that 15 percent of the tables have some sort of flaw. Smitty's Steak House, a large restaurant, orders 15 new tables from Montana Table Company.

**a.** What is the probability that at least three will be flawed?

$X$ = the number of tables flawed ~ $Bin(15, 0.15)$

(Use JMP to create the Binomial CDF table with $n = 15$ and $p = 0.15$)

$P(X \geq 3) = 1 - F(2) = 1 - 0.6042 = 0.3958$

**b.** How many tables ordered by Smitty's Steak House would you expect to have a flaw?

Since $X$ has a binomial distribution, $E(X) = n \cdot p = 15 \cdot 0.15 = 2.25$

# Chapter 3

## Section 3:

Sue's Diner orders their chairs from Montana Table Company. When Sue calls Montana Table Company to place an order, she spends 5 to 30 minutes waiting on hold where the probability of waiting on hold is constant for all minutes between 5 and 30. Suppose Sue is interested in calculating the probabilities about the length of time she waits on hold.

**a.** Find the PDF for $X$.

$X$ = length of time on hold between 5 and 30 minutes ~ $Unif(5, 30)$

$f(x) = \dfrac{1}{30 - 5} = \dfrac{1}{25}$. So we have for the PDF, $f(x) = \dfrac{1}{25}$    $5 < X < 30$

**b.** Find the probability that Sue spends between 10 and 20 minutes on hold.

$P(10 < X < 20)$ = area of the rectangle = $b \cdot h = (20 - 10) \cdot \left(\dfrac{1}{25}\right) = \dfrac{10}{25} = \dfrac{2}{5}$

## Section 4:

1. Find $Z_{0.90}$

   Find the $z$-score that has area to the right equal to 0.90. This corresponds to a $z$-score that has area in the left tail equal to 0.10. This is a special $z$-score. Thus, $Z_{0.90} = -1.282$

2. Find $Z_{0.30}$

   Find the $z$-score that has area to the right equal to 0.30. This corresponds to the $z$-score that has area in the right tail equal to 0.30. This is not a special z-score so we have to use the $N(0,1)$ CDF table. Since the $N(0,1)$ CDF table gives area to the left, we know that the $z$-score that has area to the right equal to 0.30 has area to the left equal to 0.70. Thus we look up AREA 0.70 in the $N(0,1)$ CDF table. Hence, $Z_{0.30} = 0.52$. (This is the $z$-score that has area to the left closest to 0.70)

3. The heights of cottonwood trees in Texas are normally distributed with a mean of 90 feet and standard deviation of 10 feet.

   $X$ = heights of cottonwood trees in Texas ~ $N(90, 10)$

   a. Find the probability that a randomly selected pine tree is shorter than 75 feet.

   $P(X < 75) = P(Z < -1.5) = 0.0668$

   $z = \dfrac{75 - 90}{10} = -1.5$

   b. What is the probability that a randomly selected pine tree is between 70 and 105 feet tall?

   $P(70 < X < 105) = P(-2 < X < 1.5) = 0.9332 - 0.0228 = 0.9104$

   $z = \dfrac{70 - 90}{10} = -2$ and $z = \dfrac{105 - 90}{10} = 1.5$

   c. Find the probability that a randomly selected pine tree is taller than 97 feet.

   $P(X > 97) = P(Z > 0.7) = 1 - 0.7580 = 0.2420$

   $z = \dfrac{97 - 90}{10} = 0.7$

   d. Find the height such that only 10% of trees are taller than that.

   $P(X > x) = 0.10$

   Find the specific $x$ that has area to the right equal to 10%. Thus, we need to find the $z$-score that has area in the right tail equal to 10%. This is the special

   $z$-score = 1.282. Thus $X = 90 + 1.282(10) = 102.82$ ft

# Chapter 4

## Section 2:

**1.** Suppose it is known that adolescents play an online game for an average of 81 minutes per day with standard deviation of 22 minutes. If a random sample of 40 adolescents is polled, find the. probability that the mean minutes spent playing an online game per day is less than 92 minutes.

Since $n = 40 \geq 30$, by the CLT, $\overline{X} \sim N(81, \frac{22}{\sqrt{40}})$.

$P(\overline{X} < 92) = P(Z < 3.16) = 0.9992$

$Z = \dfrac{92 - 81}{22/\sqrt{40}} = 3.16$

**2.** Suppose the life spans of exercise bicycles are known to be normally distributed with a mean of 8.25 years and a standard deviation of 3.25 years.

    **a.** What is the probability that the mean life span in a group of 10 randomly selected exercise bicycles is between 7 and 11 years?

        Since we are told $X \sim N(8.25, 3.25)$, then by the Thm, $\overline{X} \sim N(8.25, \frac{3.25}{\sqrt{10}})$

        $P(7 < \overline{X} < 11) = P(-1.22 < Z < 2.68) = 0.9963 - 0.1112 = 0.8851$

        $Z = \dfrac{7 - 8.25}{3.25 / \sqrt{10}} = -1.22$    and    $Z = \dfrac{11 - 8.25}{3.25 / \sqrt{10}} = 2.68$

    **b.** What is the probability that a randomly selected exercise bicycle has a life span of less than 13 years?

        $P(X < 13) = P(Z < 1.46) = 0.9279$

        $Z = \dfrac{13 - 8.25}{3.25} = 1.46$

## Section 3:

**1.** Suppose that 35% of all high school students take shop class. If you sample 10 high school students, what is the probability that more than 30% of the students you sample take shop class?

$P(\hat{p} > 0.30)$. But since $n \cdot p = 10 \cdot 0.35 = 3.5 < 5$, we cannot work as a $\hat{p}$ problem. We must solve it in terms of $X$ where $X \sim Bin(10, 0.35)$ and $x = n \cdot \hat{p} = 10(0.30) = 3$. Thus, we must get the Binomial CDF from JMP.

$P(\hat{p} > 0.30) = P(X > 3) = 1 - F(3) = 1 - 0.513827 = 0.4862$

**2.** Suppose that 30% of all retired people take an online class. If you sample 100 retired people, what is the probability that more than 35% of the them are taking an online class?

$P(\hat{p} > 0.35) = P(Z > 1.09) = 1 - 0.8621 = 0.1379.$

$Z = \dfrac{0.35 - 0.30}{\sqrt{\dfrac{0.30 * 0.70}{100}}} = 1.09$

# Chapter 5

## Section 3:

Suppose the administrator in charge of student tickets is interested in estimating the proportion of students who left before half time at the last football game. This administrator randomly selects 60 students who attended the last game and asks if they left before half time. Of the 60 students, 32 said yes, they did.

**a.** Find a 95% confidence interval for the true proportion of students who left before half time at the last football game.

$$95\% \text{ CI for } p: \frac{32}{60} \pm 1.96 \cdot \sqrt{\frac{\frac{32}{60} \cdot \frac{28}{60}}{60}} = (0.4071, 0.6596)$$

**b.** Can the administrator claim with 95% confidence that a majority of the students left before half time of the last game? Explain your answer.

No, the administrator cannot make that claim, since there are values contained within the CI that are less than 50%.

## Section 4:

Suppose it is believed that 36% of homes in the United States subscribe to some type of home viewing movie service. We would like to estimate the true proportion of homes in the US that subscribe to some type of home viewing movie service. What sample size should we use to ensure we estimate the truth within 3% with 95% confidence?

$$n = \left(\frac{1.96}{0.03}\right)^2 \cdot 0.36 \cdot 0.64 = 983.45. \text{ Thus } n = 984.$$

## Section 5:

**1.** Suppose it was known that people talked on their home phone for an average of 21 minutes per day. Researchers want to know if having such easy access to cell phones has changed the amount of time people talk on a phone. Thus, they would like to estimate with 95% confidence the average number of minutes that people talk on a phone (cell or land). Suppose a random sample of 40 people is polled, and the mean number of minutes talked on a land line or cell phone is 24 minutes with standard deviation 5 minutes.

**a.** Estimate with 95% confidence the average number of minutes that people talk on a phone (cell or land)

CI for $\mu$: (we will assume the population is normally distributed. Since we do not have the data we cannot verify this assumption)

$$\bar{x} \pm t_{\frac{\alpha}{2}} \cdot \frac{s}{\sqrt{n}} = 24 \pm 2.023 \cdot \frac{5}{\sqrt{40}} = (22.40, 25.60)$$

**b.** Can you be 95% confident that the average number of minutes people talk on a phone (cell or land) is more than 21 minutes? Explain your answer.

Yes, since all the values in the CI are greater than 21.

2. Suppose a city architect is interested in the heights of the buildings in his town. He would like to know if the average height of the buildings in his town is different than 105 ft. The architect randomly selects 10 buildings and their heights, in feet, are listed below.

| 106 | 109 | 107 | 108 | 112 | 113 | 104 | 105 | 115 | 114 |

**a.** Find a 99% confidence interval for the average height of the buildings.

Since we have the data, we will use JMP to find the CI.

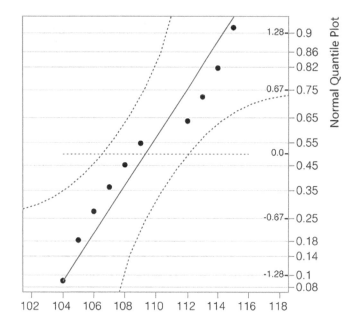

Since the data values are closely clustered to the line, we may assume that the population is normally distributed.
A 99% CI is (105.25, 113.35)

### Confidence Intervals

| Parameter | Estimate | Lower CI | Upper CI | 1-Alpha |
|-----------|----------|----------|----------|---------|
| Mean | 109.3 | 105.2453 | 113.3547 | 0.990 |
| Std Dev | 3.945462 | 2.437031 | 8.986234 | 0.990 |

**b.** Can the city architect conclude with 99% confidence that the average height is different than 105 ft.

Yes, since all the values within the CI are different than 105.

# Chapter 6

## Section 3:

The Southern Online Viewing (SOW) Company claims that 60% of online video subscribers recommend its viewing platform over that of the Northern Online Viewing (NOW) Company's platform. To test this claim against the alternative that the actual proportion is less than 60%, a random sample of 50 online viewers was taken and 52% said they prefer SOW's platform over NOW's. At the 5% significance level, what conclusion can be reached?

1. Pop = online video subscribers

    POI = $p$ = proportion of viewers that prefer SOW over NOW

2. $\alpha = 0.05$
3. Ho: $p = 0.60$   versus   Ha: $p < 0.60$
4. $Z_p$-test $\sim N(0,1)$
5. $n \cdot p = 50(0.60) = 30 > 5$   and   $n \cdot q = 50(0.40) = 20 > 5$
6. pt. est.: $\hat{p} = 0.52$. Thus our OV is $Z_p = \dfrac{0.52 - 0.60}{\sqrt{\dfrac{0.60 \cdot 0.40}{50}}} = -1.15$

7. RR: $Z_p \leq -1.645$        p-value: $P(\hat{p} \leq 0.52) = P(Z \leq -1.15) = 0.1251$

8. Fail to Reject Ho. At the 5% significance level there is not enough evidence to conclude that the proportion of customers that favor SOW over NOW is less than 60%.

## Section 4:

1. Suppose you have the following null and alternative hypothesis:   Ho: It is too cold to ski
                                                                     Ha: It is not too cold to ski

    a. In the context of this problem, describe the power of the test.

       Power = the probability of rejecting a false null hypothesis. So in the context of this problem, power is the probability of concluding that is it NOT too cold to ski when in reality it is NOT too cold to ski.

    b. In the context of this problem, describe what $\alpha$ represents.

       $\alpha$ = probability of making a type-I error, where a type-I error is rejecting a true null hypothesis. So in the context of this problem, $\alpha$ is the probability of concluding that it is NOT too cold to ski when in reality it IS too cold to ski.

    c. In the context of this problem, describe what $\beta$ represents.

       $\beta$ = probability of making a type-II error, where a type-II error is failing to reject a false null hypothesis. So in the context of this problem, $\beta$ is the probability of concluding that it IS too cold to ski when in reality it is NOT too cold to ski.

**2.** To increase power, should $n$ be increased or decreased?

Increased

**3.** To increase power, should $\alpha$ be increased or decreased?

Increased

**4.** Suppose the critical value of a two-tailed test equals $|2.064|$.

    **a.** What is the rejection region?

        $TS \leq -2.064$    or    $TS \geq 2.064$

        (where $TS$ should be the symbol used to represent the test statistic used)

    **b.** Suppose the observed value equals –3.45. What is the decision of the test?

        The decision should be Reject Ho. (Since the OV is in the RR)

**5.** Suppose the p-value equals 0.0447 and the level of significance equals 5%. What is the decision of the test?

The decision should be Reject Ho. (Since the p-value is less than $\alpha$)

## Section 5:

**1.** The Southern Online Viewing (SOW) Company would like to see if they are increasing their average length of viewing per customer over last year. The average amount of SOW content watched by a subscriber per day last year was 1.8 hours. A random sample of 68 viewers found that the average length of viewership was 1.85 hours with a standard deviation of 0.45 hours. At the 5% significance level, is there significant evidence to conclude the claim from SOW? Assume all necessary assumptions are valid.

    **1.** Pop = SOW customers        POI = $\mu$ = average length of viewing per customer

    **2.** $\alpha = 0.05$            **3.** Ho: $\mu = 1.8$    versus    Ha: $\mu > 1.8$

    **4.** $t$-test $\approx t_{(67)}$        **5.** We have to assume the population is normally distributed.

    **6.** pt. est.: $\bar{x} = 1.85$. Thus our OV is $t = \dfrac{1.85 - 1.8}{0.45 / \sqrt{68}} = 0.92$

    **7.** RR: $T \geq 1.668$      p-value: $P(\bar{x} \geq 1.85) = P(T \geq 0.92) > 0.10$

    **8.** Fail to Reject Ho. At the 5% significance level there is not enough evidence to conclude that the average length of viewing per customer is greater than 1.8 hours.

2. Suppose a researcher in public health is interested in determining if the average starting salary for those with a master's degree in public health is changing. The data below represents the starting salaries for 22 recent graduates that received a master's degree in public health. At the 1% significance level, can the researcher conclude that the average starting salary for those with a master's degree in public health is different from $55,000?

66.251   67.441   53.159   56.017   61.264   64.281   61.488   63.665   66.792   63.107   53.636

55.236   54.651   61.213   58.26   55.28   57.421   55.102   44.188   49.53   45.767   60.564

1. Pop = adults with a master's degree in public health

   POI = μ = average starting salary

2. α = 0.01                    3. Ho: μ = 55   versus   Ha: μ ≠ 55

4. $t$-test ∼ $t_{(21)}$

5. Since the data values are closely clustered to the line, we may assume the population is normally distributed.

6. pt. est.: $\bar{x}$ = 57.9233. Thus our OV is $t$ = 2.1360

7. RR: $T \leq -2.831$   or   $T \geq 2.831$          p-value: $2P(\bar{x} \geq 57.9233) = 0.0446$

8. Fail to Reject Ho. At the 1% significance level there is not enough evidence to conclude that the average starting salary for adults with a master's degree in public health is different than $55,000.

*salaries*

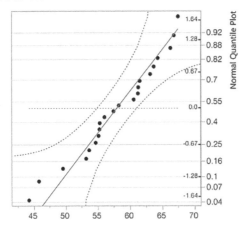

| Test Mean | |
|---|---|
| Hypothesized Value | 55 |
| Actual Estimate | 57.9233 |
| DF | 21 |
| Std Dev | 6.41928 |

| | t Test |
|---|---|
| **Test Statistic** | 2.1360 |
| Prob > ltl | 0.0446* |
| Prob > t | 0.0223* |
| Prob < t | 0.9777 |

## Section 6:

An educational researcher wants to determine if the percentage of high school students who like math is different from 50%. She takes a random sample of 85 high school students and finds that 32 of them like math.

    **a.** What is the null and alternative hypothesis that corresponds to this problem?

        Ho: $p = 0.50$   vs   Ha: $p \neq 0.50$

    **b.** Find a 95% confidence interval for the proportion of high school students who dislike math.

        Since $n \cdot p = 85(0.50) = n \cdot q = 42.5 \geq 5$, the formula is valid.

$$95\% \text{ CI for } p: \frac{32}{85} \pm 1.96 \cdot \sqrt{\frac{\frac{32}{85} \cdot \frac{53}{85}}{85}} = (0.2735, 0.4795)$$

    **c.** Based on the confidence interval found in part b, what would be the decision of the hypothesis test for part a?

        The decision is Reject Ho since the value 0.50 is not contained within the CI.

# Chapter 7

## Section 2:

An article written in Healthy Choices magazine claimed that you would gain weight once you subscribe to an online movie repository site. Data analysts for the Southern Online Viewing (SOW) Company wanted to prove the article wrong, so they surveyed 20 new subscribers and measured their weight. After two years they followed up with them to measure their weight again. The data is given below. At the 1% significance level, do the analysts have evidence to support the article's claim?

| Weight Before | 128 | 112 | 124 | 116 | 119 | 127 | 115 | 126 | 121 | 120 |
|---|---|---|---|---|---|---|---|---|---|---|
| Weight After | 129.5 | 118 | 130.5 | 118 | 128 | 133 | 120.5 | 131 | 120 | 118 |

| Weight Before | 119.5 | 128 | 115.5 | 121 | 119.5 | 117.5 | 120.5 | 118.5 | 111 | 122 |
|---|---|---|---|---|---|---|---|---|---|---|
| Weight After | 126 | 129.5 | 124 | 125 | 117 | 122 | 124.5 | 116.5 | 114.5 | 129 |

    **1.** Pop = new online movie subscribers

        POI $= \mu_d$ where $\mu_1$ = average weight before subscribing

                $\mu_2$ = average weight after subscribing

    **2.** $\alpha = 0.01$              **3.** Ho: $\mu_d = 0$   versus   Ha: $\mu_d < 0$

    **4.** $t$-test $\sim t_{(19)}$         **5.** Since the data values are closely clustered to the line we can assume that the differenced data is normally distributed.

    **6.** pt. est.: $\bar{x}_d = -3.675$. Thus our OV is $t = -4.7019$

**7.** RR: $T \le -2.539$  p-value: $P(\bar{x}_d \le -3.675) < 0.0001$

**8.** Reject Ho. At the 1% significance level there is enough evidence to conclude that the average weight of new subscribers to an online movie site does increase once they join the site.

**9.** $\mu_d < -1.69$  I am 99% confident that the average weight gained for adults after joining an online movie subscription site is more than 1.69 pounds, on average. i.e., the average weight after joining an online movie site is more than 1.69 pounds greater than the average weight after joining an online movie site, on average.

### Distributions

*d*

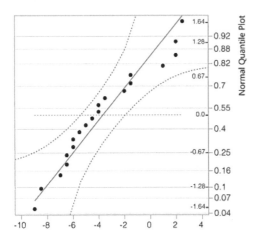

| Test Mean | |
|---|---|
| Hypothesized Value | 0 |
| Actual Estimate | -3.675 |
| DF | 19 |
| Std Dev | 3.49539 |

| | t Test |
|---|---|
| **Test Statistic** | -4.7019 |
| Prob > \|t\| | 0.0002* |
| Prob > t | 0.9999 |
| Prob < t | <.0001* |

| One-sided Confidence Interval | | | | |
|---|---|---|---|---|
| **Parameter** | **Estimate** | **Lower CI** | **Upper CI** | **1-Alpha** |
| Mean | -3.675 | . | -1.69016 | 0.990 |
| Std Dev | 3.495392 | . | 5.514837 | 0.990 |

## Section 3:

Suppose you are interested in determining if men make more money than women, on average. To test, a random sample of ten men and ten women is obtained. At the 5% significance level, can you conclude that the average salary for men is higher than it is for women? The data given below represents the salaries in thousands.

| Men | 67.441 | 53.159 | 56.017 | 61.264 | 64.281 | 61.488 | 63.665 | 66.792 | 63.107 | 53.636 |
|---|---|---|---|---|---|---|---|---|---|---|
| Women | 54.651 | 61.213 | 58.26 | 55.28 | 57.421 | 55.102 | 44.188 | 49.53 | 45.767 | 60.564 |

1. Pop = adults    POI: $\mu_2 - \mu_1$   where $\mu_1$ = average salary for men and
   $\mu_2$ = average salary for women

2. $\alpha = 0.05$    3. Ho: $\mu_2 = \mu_1$   versus   Ha: $\mu_2 < \mu_1$    4. $t_p$ – test $\sim t_{(18)}$

5. Verifying both populations are normally distributed: Since the data values are closely clustered to the line, we may assume both populations are normally distributed.

   Verifying equal population variances: Ho: $\sigma_1^2 = \sigma_2^2$ versus Ha: $\sigma_1^2 \neq \sigma_2^2$.

   Since the Brown-Forsythe p-value = 0.8149 > 0.10, we fail to reject Ho and may assume the population variances are equal.

6. $t_p = -2.78608$

7. RR: $T_p \leq -1.734$        p-value: $P((\bar{x}_2 - \bar{x}_1) \leq -6.887) = 0.0061$

8. Reject Ho. At the 5% significance level, the evidence suggest that the average salary for men is greater than the average salary for women.

9. Since Ha: $\mu_2 < \mu_1$, we just want the upper bound. $-6.887 + 1.734(2.472) = -2.6006$

Thus, CI: $\mu_2 - \mu_1 < -2.6006$.

I am 95% confident that the average salary for men is greater than the average salary for women by an amount more than $2,600.60, on average.

### Oneway Analysis of salaries By x

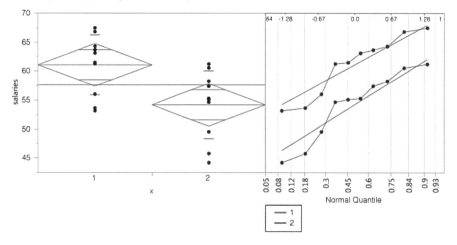

Oneway Anova

t Test

2-1

### Assuming equal variances

| Assuming equal variances | | | |
|---|---|---|---|
| Difference | -6.887 | t Ratio | -2.78608 |
| Std Err Dif | 2.472 | DF | 18 |
| Upper CL Dif | -1.694 | Prob > ltl | 0.0122* |
| Lower CL Dif | -12.081 | Prob > t | 0.9939 |
| Confidence | 0.95 | Prob < t | 0.0061* |

## Means for Oneway Anova

| Level | Number | Mean | Std Error | Lower 95% | Upper 95% |
|---|---|---|---|---|---|
| 1 | 10 | 61.0850 | 1.7480 | 57.413 | 64.757 |
| 2 | 10 | 54.1976 | 1.7480 | 50.525 | 57.870 |

Std Error uses a pooled estimate of error variance

## Tests that the Variances are Equal

| Level | Count | Std Dev | MeanAbsDif to Mean | MeanAbsDif to Median |
|---|---|---|---|---|
| 1 | 10 | 5.143670 | 4.088600 | 3.972200 |
| 2 | 10 | 5.886793 | 4.621560 | 4.350000 |

| Test | F Ratio | DFNum | DFDen | p-Value |
|---|---|---|---|---|
| O'Brien[.5] | 0.2862 | 1 | 18 | 0.5992 |
| Brown-Forsythe | 0.0564 | 1 | 18 | 0.8149 |
| Levene | 0.1510 | 1 | 18 | 0.7021 |
| Bartlett | 0.1548 | 1 | . | 0.6940 |
| F Test 2-sided | 1.3098 | 9 | 9 | 0.6942 |

# Section 4:

Health researchers are interested in determining if more men smoke than women do. A random sample of 380 men included 97 who smoke, and a random sample of 345 women included 69 who smoke. Use a 0.1 significance level to test the claim that the proportion of men who smoke is greater than the proportion of women who smoke.

**1.** Pop = adults    POI: $p_1 - p_2$ where   $p_1$ = proportion of men that smoke
$p_2$ = proportion of women that smoke

**2.** $\alpha = 0.1$      **3.**  Ho: $p_1 = p_2$   vs   Ha: $p_1 > p_2$

**4.** $n_1 \hat{p}_1 = 380 \dfrac{97}{380} = 97 \geq 5$       $n_2 \hat{p}_2 = 345 \dfrac{69}{345} = 69 \geq 5$

$n_1 \hat{q}_1 = 380 \dfrac{283}{380} = 283 \geq 5$       $n_2 \hat{q}_2 = 345 \dfrac{276}{345} = 276 \geq 5$

**5.** Since JMP put $p_2$ first, we must think of our Ha in terms of $p_2$ first. So we have Ha: $p_2 < p_1$.

So we have for the p-value: $P(\hat{p}_2 - \hat{p}_1 \leq 0) = 0.0387$

**6.** CI: Since two-tailed, the CI is testing $p_2 \neq p_1$. The CI: $(-0.1348, 0.0251)$.

**7.** Since the p-value is greater than 1% and the CI contains 0, our decision is Fail to reject Ho. Thus, at the 1% significance level, there is not enough evidence to suggest that the proportion of men that smoke is greater than the proportion of women that smoke.

*Contingency Analysis of smoking status By sex*

*Two Sample Test for Proportions*

| Description | Proportion Difference | Lower 99% | Upper 99% |
|---|---|---|---|
| P(smoker|female)-P(smoker|male) | -0.05526 | -0.13476 | 0.025134 |

| Adjusted Wald Test | Prob |
|---|---|
| P(smoker|female)-P(smoker|male) $\geq$ 0 | 0.9613 |
| P(smoker|female)-P(smoker|male) $\leq$ 0 | 0.0387* |
| P(smoker|female)-P(smoker|male) = 0 | 0.0774 |

# Chapter 8

## Section 2:

Is the student grade classification at Baylor the same as it is for other universities? In the United States, of those attending a university, the percentage of freshmen students is 25%, the percentage of sophomore students is 16%, the percentage of college juniors is 18%, the percentage of seniors is 23%, and the percentage of graduate or "other" students is 18%. For the 2012/2013 school year, Baylor had 3739 freshmen, 2743 sophomores, 2849 juniors, 3399 seniors, and 2634 graduate or "other" students. At the 1% significance level, can you conclude that the proportions for Baylor differ for the US as a whole?

1. Pop = students

   POI = $p_1$ = proportion of freshmen, $p_2$ = proportion of sophomores,
        $p_3$ = proportion of juniors, $p_4$ = proportion of juniors,
        $p_5$ = proportion of graduate/"other" students

2. $\alpha = 0.01$              **3.** Ho: $p_1 = 0.25, p_2 = 0.16, p_3 = 0.18, p_4 = 0.23, p_5 = 0.18$
                                       Ha: at least one differs from assumed

**4.** $\chi^2$- Goodness of Fit - test $\sim \chi^2_{(4)}$

**5.** $E_1 = 15{,}364(0.25) = 3{,}841 \geq 5$

$E_2 = 15{,}364(0.16) = 2{,}458.24 \geq 5$

$E_3 = 15{,}364(0.18) = 2{,}765.52 \geq 5$

$E_4 = 15{,}364(0.23) = 3{,}533.72 \geq 5$

$E_5 = 15{,}364(0.18) = 2{,}765.52 \geq 5$

**6.** $\chi^2 = 49.6057$   **7.**   P-value: $P(\chi^2 \geq 49.6057) < 0.0001$

**8.** Reject Ho

**9.** Bonferroni CI: Let $\alpha^* = 0.01/5 = 0.002$. Have JMP create five 99.8% confidence intervals.

$p_1$: (0.2328, 0.2542) Since CI contains 0.25, we cannot conclude $p_1$ is different than 25%

$p_2$: (0.1692, 0.1883) Since CI does not contain 0.16, we can conclude $p_2$ is different than 16%

$p_3$: (0.1759, 0.1953) Since CI contains 0.18, we cannot conclude $p_3$ is different than 18%

$p_4$: (0.2111, 0.2318) Since CI contains 0.23, we cannot conclude $p_4$ is different than 23%

$p_5$: (0.1622, 0.1810) Since CI contains 0.18, we cannot conclude $p_5$ is different than 18%

At the 1% significance level, the only significant difference is with the proportion of sophomores. We are 99% confident that the proportion of sophomores is between 16.92% and 18.83%, on average.

### Classification

**Test Probabilities**

| Level | Estim Prob | Hypoth Prob |
|---|---|---|
| Freshmen | 0.24336 | 0.25000 |
| Junior | 0.18543 | 0.18000 |
| Others | 0.17144 | 0.18000 |
| Seniors | 0.22123 | 0.23000 |
| Sophomores | 0.17853 | 0.16000 |

| Tests | ChiSquare | DF | Prob > Chisq |
|---|---|---|---|
| Likelihood Ratio | 48.5684 | 4 | < .0001* |
| Pearson | 49.6057 | 4 | < .0001* |

Method: Fix hypothesized values, rescale omitted          *Confidence Intervals*

| Level | Count | Prob | Lower CI | Upper CI | 1-Alpha |
|---|---|---|---|---|---|
| Freshmen | 3739 | 0.24336 | 0.232824 | 0.254217 | 0.998 |
| Junior | 2849 | 0.18543 | 0.175941 | 0.195317 | 0.998 |
| Others | 2634 | 0.17144 | 0.162248 | 0.181039 | 0.998 |
| Seniors | 3399 | 0.22123 | 0.211058 | 0.231751 | 0.998 |
| Sophomores | 2743 | 0.17853 | 0.169187 | 0.188281 | 0.998 |
| Total | 15364 | | | | |

Note: Computed using score confidence intervals.

## Section 3:

A brokerage firm wants to see whether the type of account a customer has between the choices bronze, silver, gold, platinum, or diamond is related to who makes the trade, the customer him/herself or a professional trader hired by the person. A random sample of trades made for its customers over the past year was randomly chosen and it was found that for the trades made by a hired professional broker 55 were at the bronze level, 62 were silver, 75 were gold, 72 were platinum, and 83 were diamond. Of the trades made by the customer him/herself 75 were at the bronze level, 80 were silver, 82 were gold, 65 were platinum, and 67 were diamond. At the 2.5% significance level, what conclusion can be reached?

1. Pop = customers

   WTS: If the type of account between the choices bronze, silver, gold, platinum, or diamond is related to who made the trade; a professional or the customer him/herself.

2. $\alpha = 0.025$

3. Ho: independent   vs   Ha: dependent

4. $\chi^2$ - Test of Independence - test $\sim \chi^2_{(4)}$

5. From the output, $E_i \geq 5$ for each cell.

6. $\chi^2 = 7.066$

7. $P(\chi^2 \geq 7.066) = 0.1325$

8. Fail to Reject Ho. At the 2.5% significance level, there is not enough evidence to say that the type of trade made between the choices bronze, silver, gold, platinum, and diamond is related to who made the trade; the customer him/herself or a professional.

*Contingency Analysis of type By made by*
*Contingency Table*
made by By type

| Count Expected | Bronze | Diamond | Gold | Platinum | Silver | Total |
|---|---|---|---|---|---|---|
| **Customer** | 75 66.9972 | 67 77.3045 | 82 80.912 | 65 70.6047 | 80 73.1816 | 369 |
| **Professional** | 55 63.0028 | 83 72.6955 | 75 76.088 | 72 66.3953 | 62 68.8184 | 347 |
| **Total** | 130 | 150 | 157 | 137 | 142 | 716 |

Tests

| N | DF | -LogLike | RSquare (U) |
|---|---|---|---|
| 716 | 6 | 3.5404383 | 0.0031 |

| Tests | ChiSquare | Prob>ChiSq |
|---|---|---|
| Likelihood Ratio | 7.081 | 0.1317 |
| Pearson | 7.066 | 0.1325 |

# Section 4:

Campus Life administrators want to determine if students are similar with respect to if and where they work. A random sample of freshman found that 320 do not work, 260 work as a student worker for the University, and 450 work off-campus. A random sample of sophomores found that 280 do not work, 220 work as a student worker for the University, and 350 work off-campus. A random sample of juniors found that 250 do not work, 315 work as a student worker for the University, and 380 work off-campus. A random sample of seniors found that 280 do not work, 220 work as a student worker for the University, and 350 work off-campus. At the 5% significance level what conclusion can be reached?

1. Pop = students

   WTS: If the students based on classification are similar with respect to work status.

2. $\alpha = 0.05$

3. Ho: homogeneous   vs   Ha: heterogeneous

4. $\chi^2$ - Test of Homogeneity - test $\approx \chi^2_{(6)}$

5. From the output, $E_i \geq 5$ for each cell.

6. $\chi^2 = 25.082$

7. $P(\chi^2 \geq 25.082) = 0.0003$

8. Reject Ho. At the 5% significance level, the evidence suggests that freshmen, sophomores, juniors, and seniors are not similar with respect to work status.

## Contingency Analysis of work status By classification

Contingency Table

| Count<br>Expected | No | Off campus | On campus | Total |
|---|---|---|---|---|
| **Fr** | 320<br>316.707 | 450<br>428.816 | 260<br>284.476 | 1030 |
| **Ju** | 250<br>290.571 | 380<br>393.429 | 315<br>261 | 945 |
| **Se** | 280<br>261.361 | 350<br>353.878 | 220<br>234.762 | 850 |
| **So** | 280<br>261.361 | 350<br>353.878 | 220<br>234.762 | 850 |
| **Total** | 1130 | 1530 | 1015 | 3675 |

Tests

| N | DF | -LogLike | RSquare (U) |
|---|---|---|---|
| 3675 | 6 | 12.348754 | 0.0031 |

| Tests | ChiSquare | Prob>ChiSq |
|---|---|---|
| Likelihood Ratio | 24.698 | 0.0004* |
| Pearson | 25.082 | 0.0003* |

# Chapter 9

## Section 1:

With the increase in health insurance cost, many healthy Americans are looking into different health-care plans. Americans can choose from among four different health plans. It is of interest to know whether the mean yearly co-pay cost is different for the four different health plans for healthy Americans. To test, four random samples of healthy Americans were selected, where each sample represents one of the health plans, and the yearly co-pay cost is listed below. At the 5% significance level, what conclusion can be reached?

| Plan A: | 1198 | 1195 | 1194 | 1203 | 1204 | 1205 | 1206 | 1199 | 1207 | 1200 |
| Plan B: | 1227 | 1222 | 1234 | 1218 | 1225 | 1227 | 1228 | 1226 | 1212 | 1229 |
| Plan C: | 1228 | 1230 | 1239 | 1237 | 1233 | 1234 | 1238 | 1232 | 1229 | 1218 |
| Plan D: | 1216 | 1205 | 1212 | 1214 | 1211 | 1205 | 1199 | 1202 | 1208 | 1203 |

**Hypothesis:**

Ho: $\mu_A = \mu_B = \mu_C = \mu_D$      or      Ho: $\tau_j = 0$ *for all j*

Ha: not all means are equal            Ha: $\tau_j \neq 0$ *for some j*

**Observed Value:** $F = 64.6517$

**P-value:** $P(F > 64.6517) < 0.001$

**Decision:** Reject Ho

**If appropriate, use Tukey's Test to determine which means differ:**

$\mu_A$ VS $\mu_B$   +   →   The means are significantly different

$\mu_A$ VS $\mu_C$   +   →   The means are significantly different

$\mu_A$ VS $\mu_D$   −   →   The means are not significantly different

$\mu_B$ VS $\mu_C$   +   →   The means are significantly different

$\mu_B$ VS $\mu_D$   +   →   The means are significantly different

$\mu_C$ VS $\mu_D$   +   →   The means are significantly different

**Conclusion:** At the 5% significance level, the only significant difference in the mean yearly co-pay cost for the four health plans is for Plan A and Plan D.

*Oneway Analysis of Column 1 By x*

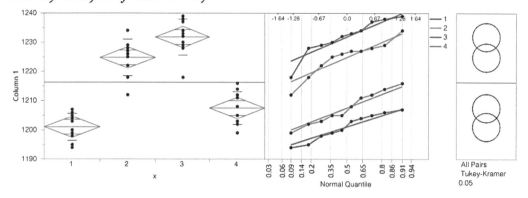

## Oneway Anova

### Analysis of Variance

| Source | DF | Sums of Squares | Mean Square | F-ratio | Prob > F |
|---|---|---|---|---|---|
| x | 3 | 6209.8000 | 2069.93 | 64.6517 | <.0001* |
| Error | 36 | 1152.6000 | 32.02 | | |
| C. Total | 39 | 7362.4000 | | | |

### Tests that the Variances are Equal

| Level | Count | Std Dev | MeanAbsDif to Mean | MeanAbsDif to Median |
|---|---|---|---|---|
| 1 | 10 | 4.581363 | 3.900000 | 3.900000 |
| 2 | 10 | 6.160808 | 4.480000 | 4.200000 |
| 3 | 10 | 6.142746 | 4.440000 | 4.400000 |
| 4 | 10 | 5.602579 | 4.700000 | 4.700000 |

| Test | F Ratio | DFNum | DFDen | p-Value |
|---|---|---|---|---|
| O'Brien[.5] | 0.2567 | 3 | 36 | 0.8561 |
| Brown-Forsythe | 0.0897 | 3 | 36 | 0.9653 |
| Levene | 0.1089 | 3 | 36 | 0.9544 |
| Bartlett | 0.3086 | 3 | . | 0.8192 |

### Means Comparisons

### Comparisons for all pairs using Tukey-Kramer HSD

### Confidence Quantile

| $q^*$ | Alpha |
|---|---|
| 2.69323 | 0.05 |

### HSD Threshold Matrix

| Abs(Dif)-HSD | 3 | 2 | 4 | 1 |
|---|---|---|---|---|
| 3 | -6.815 | 0.185 | 17.485 | 23.885 |
| 2 | 0.185 | -6.815 | 10.485 | 16.885 |
| 4 | 17.485 | 10.485 | -6.815 | -0.415 |
| 1 | 23.885 | 16.885 | -0.415 | -6.815 |

Positive values show pairs of means that are significantly different.

# Section 2:

Due to the increase in monthly healthcare premiums, many healthy Americans are investigating to see which health care system has the lowest monthly premiums. The premiums are determined by the number of members in the family. The data below is the monthly premiums for families for four different health plans.

|  | Plan A | | Plan B | | Plan C | | Plan D | |
|---|---|---|---|---|---|---|---|---|
| 2 Family Members | 1150 | 1145 | 1165 | 1154 | 1121 | 1133 | 1159 | 1163 |
| 3–5 Family Members | 1345 | 1401 | 1420 | 1505 | 1431 | 1401 | 1390 | 1436 |
| 6+ Family Members | 1522 | 1602 | 1599 | 1654 | 1642 | 1603 | 1684 | 1699 |

**a.** At the 2.5% significance level, is there a difference in the average monthly premiums for the four different health plans.

**Hypothesis:**

$Ho: \mu_A = \mu_B = \mu_C = \mu_D$  or  $Ho: \tau_j = 0 \text{ for all } j$

Ha: not all means are equal   $Ha: \tau_j \neq 0 \text{ for some } j$

**Observed Value:** $F = 3.456$

**P-value:** $P(F > 3.456) = 0.0381$

**Decision:** Fail to Reject Ho

**If appropriate, use Tukey's Test to determine which means differ:** Since we failed to reject Ho, there is no need to do Tukey's test.

**Conclusion:** At the 2.5% significance level, there is not enough evidence to conclude that there is a difference in the mean monthly premium for the four healthcare plans.

**b.** Test to determine if there is a benefit to blocking based on family size.

**Hypothesis:**

$Ho: \beta_j = 0 \text{ for all } j$

$Ha: \beta_j \neq 0 \text{ for some } j$

**Observed Value:** $F = 334.981$

**P-value:** $P(F > 334.981) < 0.0001$

**Decision:** Reject Ho

**Conclusion:** There was a benefit to blocking on family size. (Family size is a confounding factor)

## Oneway Analysis of premium By Plan

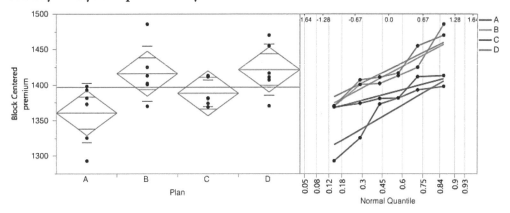

## Oneway Anova
## Analysis of Variance

| Source | DF | Sums of Squares | Mean Square | F ratio | Prob > F |
|---|---|---|---|---|---|
| Plan | 3 | 14185.33 | 4728 | 3.4656 | 0.0381* |
| # in Family | 2 | 914105.08 | 457053 | 334.9881 | <.0001* |
| Error | 18 | 24558.92 | 1364 | | |
| C. Total | 23 | 952849.33 | | | |

## Tests that the Variances are Equal

| Level | Count | Std Dev | MeanAbsDif to Mean | MeanAbsDif to Median |
|---|---|---|---|---|
| A | 6 | 42.00818 | 34.25000 | 30.12500 |
| B | 6 | 38.66981 | 26.19444 | 25.16667 |
| C | 6 | 19.13907 | 15.97222 | 13.70833 |
| D | 6 | 35.85300 | 27.25000 | 25.54167 |

| Test | F Ratio | DFNum | DFDen | Prob > F |
|---|---|---|---|---|
| O'Brien[.5] | 0.6329 | 3 | 20 | 0.6024 |
| Brown-Forsythe | 0.4494 | 3 | 20 | 0.7205 |
| Levene | 0.9178 | 3 | 20 | 0.4502 |
| Bartlett | 0.9296 | 3 | . | 0.4254 |

# Chapter 10

## Section 3:

A pharmaceutical company has developed a new drug for patients with Wolfe-Parkinson-White syndrome designed to control paroxysmal supraventricular tachycardia (PSVT). Patients with PSVT have a heart rate of more than 100 beats per minute. (BPM) Suppose you are interested in modeling the relationship between the decrease in pulse rate in BPM and dosage in cc. Different doses were administered to 16 randomly selected patients and 20 minutes later the decrease in each patient's pulse rate was recorded. The data collected is given below.

| Dosage | 2 | 4 | 1.5 | 1 | 3 | 3.5 | 2.5 | 3 | 3.5 | 1.5 | 2 | 2 | 3.75 | 3.5 | 1.75 | 1.25 |
|---|---|---|---|---|---|---|---|---|---|---|---|---|---|---|---|---|
| Drop in HR | 14 | 24 | 11 | 6 | 18 | 22 | 17 | 15 | 22 | 9 | 16 | 14 | 23 | 25 | 12 | 12 |

   **a.** Which variable is the independent variable?   Dosage

   **b.** Which variable is the dependent variable?   Drop in heart rate

   **c.** Is a linear model appropriate? Explain your answer.

      Yes, since the data values are forming a linear shape.

   **d.** If a linear model is appropriate, find the line of best fit.

      $\hat{y} = 2.5983762 + 5.4949932x$

   **e.** If appropriate, interpret the coefficients of the model.

      $b_0 = 2.5984$   The $y$-intercept does not have any interpretation since we did not consider values near zero.

      $b_1 = 5.4950$   If we increase the dosage by one cc, the average drop in heart rate will increase by 5.4950 beats per minute.

   **f.** If appropriate, estimate the drop in pulse rate for a dosage level of 2.75 cc. Interpret the estimate found.

      $\hat{y} = 2.5983762 + 5.4949932(2.75) = 17.71$

      We estimate that anyone who receives a dosage of 2.75 cc will have an average drop in pulse rate of 17.71 BPM.

   **g.** If appropriate, estimate the drop in pulse rate for a dosage level of 5.5 cc.

      Since we did not consider values of $X$ that big, it would be extrapolating too far beyond the range of $X$ values considered and thus would not be appropriate to use the model to estimate for $X = 5.5$.

*Bivariate Fit of y By x*

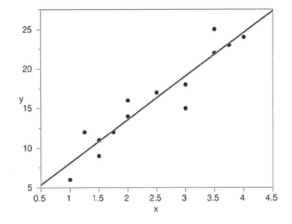

---
□ —— Linear Fit
---

| **Linear Fit** |
| --- |
| y = 2.5983762 − 5.4949932*x |

| **Summary of Fit** | |
| --- | --- |
| RSquare | 0.898601 |
| RSquare Adj | 0.891359 |
| Root Mean Square Error | 1.87423 |
| Mean of Response | 16.25 |
| Observations (or Sum Wgts) | 16 |

| **Parameter Estimates** | | | | |
| --- | --- | --- | --- | --- |
| Term | Estimate | Std Error | t Ratio | Prob>ltl |
| Intercept | 2.5983762 | 1.312124 | 1.98 | 0.0677 |
| x | 5.4949932 | 0.493328 | 11.14 | <.0001* |

## Section 4:

(Using the same information from Section 3 problem)

**a.** Find and interpret a 99% confidence interval for the slope of the regression line.

99% CI for $\beta_1$:  5.4950 ± 2.977(0.4933) = (4.03, 6.96).

We are 99% confident that if we increase the dosage by one cc, the average drop in pulse rate will increase between 4.03 and 6.96 BPM, on average.

| **Parameter Estimates** | | | | |
| --- | --- | --- | --- | --- |
| Term | Estimate | Std Error | t Ratio | Prob>ltl |
| Intercept | 2.5983762 | 1.312124 | 1.98 | 0.0677 |
| x | 5.4949932 | 0.493328 | 11.14 | <.0001* |

**b.** Find and interpret a 99% confidence interval for the average drop in pulse rate for a dosage level of 2.75 cc.

99% CI for $\mu_{Y|x}$: (16.26, 19.16)

We are 99% confident that anyone who receives a dosage of 2.75 ccs will have an average drop in pulse rate between 16.26 and 19.16 BPM, on average.

(Use JMP to get CI. Make sure and define $\alpha = 0.01$)

| | | | | | | | |
|---|---|---|---|---|---|---|---|
| 1.25 | | 12 | 7.179847771 | 11.754387682 | 3.4371867169 | 15.497048736 | |
| 2.75 | | • | 16.261264998 | 19.157950158 | 11.94539381 | 23.473821346 | |

**c.** Find and interpret a 99% prediction interval for a dosage level of 2.75 cc.

PI: (11.95, 23.47)

We are 99% confident that this specific person who received a dosage of 2.75 cc will have a drop in pulse rate between 11.95 and 23.47 BPM, on average.

| | | | | | | | |
|---|---|---|---|---|---|---|---|
| 1.25 | | 12 | 7.179847771 | 11.754387682 | 3.4371867169 | 15.497048736 | |
| 2.75 | | • | 16.261264998 | 19.157950158 | 11.94539381 | 23.473821346 | |

(Use JMP to get PI. Make sure and define $\alpha = 0.01$)

## Section 5:

(Using the same information from Section 3 problem)

**a.** Find the correlation coefficient. Is there a strong linear relationship? Is it a direct or an indirect?

From JMP we get $r^2 = 0.898601$. Thus, $r = \sqrt{0.898601} = 0.9479$.

($r$ is positive, since $b_1$ is positive.) Since r is close to 1, there is a strong direct linear relationship between dosage level and drop in pulse rate.

| *Summary of Fit* | |
|---|---|
| RSquare | 0.898601 |
| RSquare Adj | 0.891359 |
| Root Mean Square Error | 1.87423 |
| Mean of Response | 16.25 |
| Observations (or Sum Wgts) | 16 |

**b.** Find the coefficient of determination.

From JMP we get $r^2 = 0.8986$.

**c.** Interpret the coefficient of determination using the variance interpretation.

Approximately 89.86% of the variation in drop in pulse rate can be explained by the linear relationship between dosage level and drop in pulse rate.

**d.** Interpret the coefficient of determination using the error interpretation.

We reduce our prediction error by approximately 89.86% using the linear regression line to estimate drop in pulse rate versus just using the average pulse rate to estimate drop in pulse rate.

## Section 6:

(Using the same information from Section 3 problem)

Perform residual analysis to verify the assumptions of the model.

Residual Analysis consists of three steps:

**1.** Create Residual Plot:

***Diagnostics Plots***
***Residual by Predicted Plot***

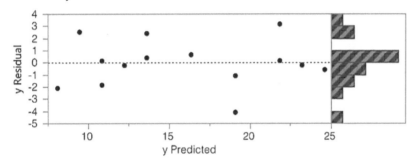

Since the residuals are randomly distributed around zero, i.e., there is no pattern, we can assume the following to be valid.

**1.** The relationship between dosage level and average drop in pulse rate is linear.

**2.** The errors are independent.

**3.** The errors have constant variance and zero mean.

**2.** QQ-plot of the residuals.

*Residual Normal Quantile Plot*

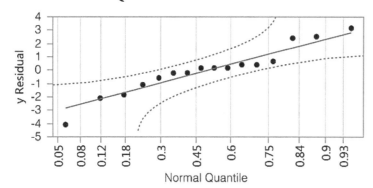

Since the residuals are closely clustered to the line, we may assume that they are normally distributed.

**3.** Check for Outliers.

$\pm\,3(S) = \pm\,3(\text{Root Mean Square Error}) = \pm\,3(1.87432) = \pm\,5.62$

**The residuals found from JMP:**

| | | | |
|---|---|---|---|
| 0.4116373478 | −0.57834912 | 0.1591339648 | −2.093369418 |
| −1.083355886 | 0.1691474966 | 0.6641407307 | −4.083355886 |
| 0.1691474966 | −1.840866035 | 2.4116373478 | 0.4116373478 |
| −0.204600812 | 3.1691474966 | −0.214614344 | 2.5328822733 |

Since none of the residuals are larger in magnitude to 5.62, we can conclude that there are not any outliers.